Global Climatic Catastrophes

M.I. Budyko G.S. Golitsyn Y.A. Izrael

Global Climatic Catastrophes

Translated by V.G. Yanuta

With 22 Illustrations

Springer-Verlag
Berlin Heidelberg New York
London Paris Tokyo

Michael I. Budyko
The State Hydrological Institute
Leningrad 199053, USSR

Georgi S. Golitsyn
The Institute of Atmospheric Physics
Moscow 100917, USSR

Yuri A. Izrael
State Committee on Hydrometeorology
  and Control of Natural Environment
(Goskomgidromet of the USSR)
Moscow D-376, USSR 123376

V. G. Yanuta (*Translator*)
The State Hydrological Institute
Leningrad 199053, USSR

Global Climatic Catastrophes by Michael I. Budyko, Georgi S. Golitsyn, and Yuri A. Izrael was originally published in Russian by Gidrometeoizdat, Leningrad, USSR, 1986; translated with permission.

Library of Congress Cataloging-in-Publication Data
Budyko, M. I. (Mikhail Ivanovich)
[Global'nye klimaticheskie katastrofy. English]
Global climatic catastrophes / M.I. Budyko, G.S. Golitsyn, Y.A. Izrael.
p.   cm.
Translation of: Global'nye klimaticheskie katastrofy.
Bibliography: p.
ISBN 0-387-18647-6 (U.S.)
1. Climatic changes. 2. Weather—Effect of volcanoes on.
3. Aerosols—Environmental aspects. 4. Nuclear winter.
I. Golitsyn, G. S. (Georgiĭ Sergeevich) II. Izraėl', IU. A. (IUiĭ
Antonievich) III. Title
QC981.8.C5B7913   1988
363.3'492—dc19   88-4950

© 1988 by Springer-Verlag New York Inc.
All rights reserved. This work may not be translated or copied in whole or in part without the written permission of the publisher (Springer-Verlag, 175 Fifth Avenue, New York, NY 10010, USA), except for brief excerpts in connection with reviews or scholarly analysis. Use in connection with any form of information storage and retrieval, electronic adaptation, computer software, or by similar or dissimilar methodology now known or hereafter developed is forbidden.
The use of general descriptive names, trade names, trademarks, etc. in this publication, even if the former are not especially identified, is not to be taken as a sign that such names, as understood by the Trade Marks and Merchandise Marks Act, may accordingly be used freely by anyone.

Typeset by Publishers Service, Bozeman, Montana.
Printed and bound by R.R. Donnelley & Sons, Harrisonburg, Virginia.
Printed in the United States of America.

9 8 7 6 5 4 3 2 1

ISBN 3-540-18647-6 Springer-Verlag Berlin Heidelberg New York
ISBN 0-387-18647-6 Springer-Verlag New York Berlin Heidelberg

Preface

The necessity of eliminating the possibility of a large-scale nuclear conflict from the future of mankind is the most important problem of our times. There is no doubt that the probable aftereffects of such a conflict would by many times exceed the damage caused by the First and Second World Wars, the greatest in history. The question of the scale of the damage that would be inflicted upon living nature by nuclear weapons has, however, not yet been fully clarified.

It is clear that this damage would not be local, i.e., restricted to destruction in only the places of nuclear explosion. As a result of nuclear detonations, the atmosphere and hydrosphere would receive many harmful substances, including the radioactive waste products of nuclear reactions. These substances can be transferred by air flows and water currents over long distances, thus considerably increasing the area of harmful influence of nuclear bursts.

There is no doubt that the indirect effects of nuclear warfare would inflict enormous damage on mankind, since the present human society can only exist by a complicated system involving the production of foodstuffs, manufactured goods, medical supplies, etc. The destruction of even separate but important links of this system would bring about starvation, epidemics, and other calamities, which would spread to areas not directly involved in the nuclear conflict.

Proceeding from these considerations, the conclusion has repeatedly been drawn that the probable number of victims of nuclear warfare would be much greater than those killed by nuclear bursts, and would comprise a considerable part of the Earth's population.

Studies carried out recently have revealed the real danger of much greater consequences of large-scale nuclear warfare, which would be of a global nature, i.e., would apply to the biosphere as a whole. These consequences could result in the extermination of mankind and could possibly threaten the existence of the biosphere itself. The major cause of the global catastrophe that could arise from nuclear warfare is a comparatively short-term but considerable climatic change induced by many different factors, among which is a sharp increase in the number of atmospheric aerosol particles.

The theory of the initiation of global catastrophes, both in the present and in the geological past, by the climatic effects of certain external factors arose in the late 1960s, when a new scientific discipline—physical climatology—began to develop. Since this development has to a significant degree been due to the efforts of Soviet scientists, many of the new inferences concerning the physical mechanism of climate genesis and climate change were researched in the USSR much earlier than in other countries. In particular, the theory of aerosol climatic catastrophe appeared in studies by Soviet scientists more than 10 years earlier than similar conclusions in other countries.

The conclusion that climate would be catastrophically changed after large-scale nuclear warfare was also first drawn in the USSR. This conclusion was discussed at a number of national and international meetings in the early 1980s just before the appearance of the first publications on this problem. Widespread discussion in the press of the catastrophic changes of climate after nuclear warfare appeared only in 1983–84. The fact that the majority of investigators taking part in this discussion came to the conclusion that large-scale nuclear warfare can result in a climatic catastrophe deserves attention.

In this book, the causes of climatic change in the present and in the geological past are considered first to explain the physical mechanism of global climatic catastrophes. Information is then given about local and global climatic catastrophes caused by natural factors, and the possible influence of these catastrophes on the mass extinction of organisms in the geological past. Further, climatic changes due to man's activities, including a global climatic catastrophe induced by a large-scale nuclear conflict, are considered.

The results presented in this book can be used to estimate the sensitivity of the biosphere to such an extremely dangerous impact of man as nuclear warfare. The broad dissemination of information on the ecological consequences of using nuclear weapons can lower the probability of starting a nuclear conflict and possibly even banish the nuclear threat completely from the future of mankind.

Leningrad, USSR	M.I. Budyko
Moscow, USSR	G.S. Golitsyn
Moscow, USSR	Y.A. Izrael

Contents

Preface v

1 Natural Climatic Catastrophes 1
   1.1 Climatic Change 1
   1.2 Catastrophic Climatic Changes 9
   1.3 Critical Epochs of Geological History 26

2 Climatic Effects of a Nuclear Conflict 39
   2.1 Climatic Aerosol Catastrophe 39
   2.2 Other Atmospheric Effects of a Nuclear War 76
   2.3 The Reliability of the Results 79

References 87

Index 97

1
Natural Climatic Catastrophes

1.1 Climatic Change

CLIMATES OF THE GEOLOGICAL PAST

Factual information about changes of climatic conditions in the past is available basically for the last 570 million years, i.e., for the interval of time called the Phanerozoic. From the beginning of the Phanerozoic, animals with various hard tissues (skeleton forms) have been widely distributed, and their remains well preserved in sedimentary rocks. In the Phanerozoic, vegetation cover appeared also on the continents, remains of which are frequently found in sediments. As a result, a comprehensive study of the change in successive faunas and floras throughout the Phanerozoic appeared to be possible.

Until recently, information about past climates has been obtained by indirect methods: from data on the distribution of different animals and plants, lithogenesis, structure of the Earth's surface on which the traces of ancient glaciations were found, and so on. Today for this purpose we use paleotemperature data obtained by analyzing the isotope composition of the remains of ancient organisms, calculations based on climate theory methods and on estimates of the chemical composition of the ancient atmospheres. In particular, data on variations in the Earth's average surface air temperature for 28 Phanerozoic epochs have been thus obtained (Fig. 1; Budyko et al. 1985). As can be seen from Figure 1, average temperatures for the indicated epochs varied within the 10°C range.

The data in Figure 1 represent an average temperature for the epochs with duration from several million up to several tens of millions of years. It is evident that the range of average temperature variations for shorter time periods will be greater than the value given. In particular, throughout the Phanerozoic, two time intervals are recorded during which glaciation advanced over a considerable part of the Earth's surface. The first glaciation took place at the end of the Carboniferous and the beginning of the Permian periods, and the second was greatest in the Quaternary period. The mean surface air temperature during the first glaciation is difficult to estimate; in the second extensive glaciation, although the relevant data on mean temperatures are not presented in Figure 1 (which is difficult to do because the Quaternary continued for about 2 million years, during which the temperature varied widely), for several epochs of the Quaternary these data

2   1. Natural Climatic Catastrophes

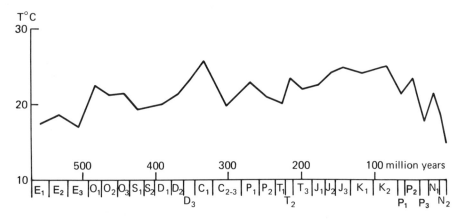

FIGURE 1. Mean surface air temperature changes (T) in the Phanerozoic.

are available. In particular, from paleotemperature data, it is seen that during the last glaciation (about 20,000 years ago) the mean surface air temperature was 5°C lower than the modern one (CLIMAP Project Members 1976). Since the present air temperature is lower than the average temperature recorded throughout all geological epochs, data for which are shown in Figure 1, it can be inferred that the range of variations in global mean air temperatures during the Phanerozoic was not less than 15°C.

There are grounds to believe that the actual range of the Phanerozoic mean temperature change for long-term time periods (exceeding 1,000 years) was not considerably greater than this value. This is partly confirmed by paleotemperature data. Thus, for example, Frakes (1979) concluded that in the middle of the Cretaceous the mean air temperature was 10–15°C higher than at present, this period being associated with one of the warmest climates during the Phanerozoic. The sum of the indicated increase and the greatest decrease in temperature during the Würm glaciation is equal to 15–20°C.

At the same time, it can be supposed that a mean temperature change in a wider range would threaten the existence of the biosphere. Even at the present relatively low mean air temperature, in a number of continental tropical regions the air temperature sometimes reaches 40–45°C. In this case, the temperature of the Earth's surface illuminated by the Sun and not covered with vegetation usually reaches far higher values.

These high temperatures are not favorable for life for the majority of animals and plants. Under these conditions, photosynthetic productivity is noticeably reduced. Many animals in tropical countries seek to conceal themselves during the hottest days in bodies of water, burrows, or in the branches of trees, which, however, does not always save them from death.

The results of experiments with crocodiles (Colbert et al. 1946) are interesting. These authors demonstrated how crocodiles, tethered on the shore of a river in the daytime with no possibility of hiding in the water, comparatively soon die from overheating. This is of interest because crocodiles, of all modern animals, are most closely related to the large reptiles of the Mesozoic which lived under the warmest climate conditions of the entire Phanerozoic.

To explain the possibility of the existence of Mesozoic reptiles in the tropics, it is of considerable importance to take into account the known feature to changing climate discovered in a number of theoretical and empirical studies (ref. Budyko 1984). With increasing global mean temperature, the greatest warming occurs in the mid- and particularly in the high latitudes, whereas in low latitudes the temperature rises comparatively little. In this connection, the maximum temperature in the tropics during the warmest epochs of the past was slightly above the modern one, although even this relatively small increase in temperature seems to present certain difficulties for animal and plant life. It might be thought that, based on this feature, an increase in temperature, noticeable if compared with the maximum for the Phanerozoic, would lead to the extinction of many living organisms.

There are also grounds to believe that during the Phanerozoic, epochs with global mean temperature noticeably lower than the temperature indicated for the Würm glaciation did not exist. This inference can be drawn from modern studies on the physical mechanism of climatic change in the geological past. The causes of variations in global mean surface air temperature are clear from Table 1, based on a recently published study by Budyko et al. (1985).

Table 1 presents estimates of variations in the mean surface air temperature (ΔT) for various geological epochs as compared with modern temperature. The value of ΔT is equal to the sum of three values: ΔT_{CO_2}, a temperature change due to variations in atmospheric carbon dioxide concentration, ΔT_s, the temperature effects of increasing solar constant due to the growing luminosity of the Sun, and ΔT_α, a temperature change due to increased or decreased albedo of the Earth.

As can be seen from Table 1, the mean air temperature varied in the geological past mainly due to changes in atmospheric carbon dioxide, which resulted in corresponding alterations of the greenhouse effect. It is found that the temperature of the lower atmosphere is a logarithmic function of atmospheric carbon dioxide concentration. It increases by about 3°C with doubling CO_2 concentration. At the same time, the temperature is affected by increasing solar radiation and varied albedo of the Earth's surface that depends on the relative area of continents and oceans, areas of various continental vegetation zones and the presence of snow and ice cover on the Earth's surface. A 1% change in solar radiation increases or decreases the surface air temperature by approximately 1.5°C; a 0.01 change in the albedo of the Earth-atmosphere system because of varying mean albedo of the Earth's surface results in increasing or decreasing the temperature by about 2°C.

Snow and ice cover can exert a pronounced effect on temperature. It is easy to comprehend that the area of snow and ice cover, after its formation, in turn depends on the thermal regime of the atmosphere, this dependence being a positive feedback by nature. Snow and ice form a permanent cover on the continents and oceans where the Earth's surface temperature is below the freezing point of water. Snow and ice cover considerably raise the Earth's surface albedo, which contributes to decreasing the absorbed solar radiation and causes further reduction in air temperature and expansion of the area covered with snow and ice. Including dependence in the climate model showed that a comparatively small decrease of global mean air temperature can appear to be sufficient for complete glaciation of the Earth ("white Earth"). This state can be achieved either by a several percent decrease of solar constant or because of an attenuation of the greenhouse effect due to the manifold reduction of the carbon dioxide content of the modern atmosphere (Budyko 1968 etc.).

As can be seen from the data of Table 1, in the beginning of the Phanerozoic, the Sun's luminosity was almost 3% less than its modern value. Earlier, in the Precambrian, in particular in the Early Precambrian, this difference was far greater, and, apparently in the earliest history of the Earth, 4.5 billion years ago, it reached 25%. Nevertheless, throughout a greater part of the Phanerozoic, the mean air temperature exceeded that of today, and even if glaciations appeared in some epochs of the Precambrian, they never enveloped the entire Earth.

This inference does not contradict the possibility of the formation of "white Earth" that proceeds from the above calculations. It has been comparatively recently established that in the Solar System there are a number of celestial bodies whose surfaces are entirely covered with ice (some satellites of Jupiter, and individual asteroids) (Marov 1981). Relatively warm climates of the remote past seem to be attributed to a far greater amount of carbon dioxide in the ancient atmosphere compared with the modern, which appreciably increased the greenhouse effect. As found in several studies, in the Precambrian, carbon dioxide content was 10 to 100 times above the modern, due to a far higher rate of influx of this gas to the atmosphere from the Earth's depths, strongly heated by the decay of long-lived radioactive elements (Budyko et al. 1985). The influence of this factor over-compensated for the impact of decreased solar radiation influx on the thermal conditions of the atmosphere, which eliminated the possibility of a global glacial advance.

As a result of this, the Earth's biosphere, i.e., the region inhabited by living organisms, could exist for over almost 4 billion years. The theory has repeatedly been proposed that the probability of this long existence of the biosphere on a planet like the Earth is extremely low. One of the reasons for this is the narrowness of temperature range, compared with the temperature variability on different celestial bodies, within which a very great number of organisms can survive.

In other words, at any time in the entire history of the Earth's biosphere, it could have been destroyed as a result of the uncoordinated change in two independent processes: the growth of the Sun's luminosity and a gradual retardation of the Earth's degassing at depth due to a decreasing mass of long-lived radioactive isotopes of some of the elements heating the depth.

This lack of coordination occurred during the last 100 million years, when as a result of a sharp decrease in degassing, the carbon dioxide mass in the atmosphere declined, which lowered the mean temperature of the surface atmospheric

TABLE 1. Mean air temperature variations in the Phanerozoic (°C)

Time Interval	Beginning and End of the Interval (ma)	ΔT_{CO_2}	$-\Delta T_s$	ΔT_α	ΔTm
Early Cambrian	570–545	3.3	3.9	3.6	3.0
Middle Cambrian	545–520	4.6	3.6	3.0	4.0
Late Cambrian	520–490	2.8	3.5	3.0	2.3
Early Ordovician	490–475	7.7	3.4	3.4	7.7
Middle Ordovician	475–450	6.3	3.2	3.4	6.5
Late Ordovician	450–435	6.3˙	3.1	3.4	6.6
Early Silurian	435–415	4.6	2.9	3.0	4.7
Late Silurian	415–402	4.6	2.8	3.0	4.8
Early Devonian	402–378	4.6	2.7	3.2	5.1
Middle Devonian	378–362	6.1	2.5	3.0	6.6
Late Devonian	362–346	7.8	2.5	3.0	8.3
Early Carboniferous	346–322	10.0	2.2	3.2	11.0
Middle-Late Carboniferous	322–282	6.1	2.1	1.0	5.0
Early Permian	282–257	9.2	1.8	0.8	8.2
Late Permian	257–236	4.7	1.7	3.2	6.2
Early Triassic	236–221	4.8	1.5	3.0	6.3
Middle Triassic	221–211	7.4	1.5	2.8	8.7
Late Triassic	211–186	5.7	1.4	2.8	7.1
Early Jurassic	186–168	6.0	1.3	3.0	7.7
Middle Jurassic	168–153	7.2	1.1	3.2	9.3
Late Jurassic	153–133	8.9	1.0	3.0	10.9
Early Cretaceous	133–101	6.9	0.8	3.0	9.1
Late Cretaceous	101–67	7.7	0.6	3.2	10.3
Paleocene	67–58	4.0	0.4	3.0	6.6
Eocene	58–37	6.0	0.3	2.8	8.5
Oligocene	37–25	0.3	0.3	2.8	2.8
Miocene	25–9	4.0	0.1	2.4	6.3
Pliocene	9–2	1.8	0	1.6	3.4

layer. The preservation of this tendency over several million years could have resulted in the complete glaciation of the Earth, that is, the destruction of the biosphere.

Thus throughout all the epochs after the formation of the biosphere, the possibility of a global ecological crisis existed, i.e., such a climatic change as would have led to the total destruction of the biosphere or (with a less significant climatic change) to the extinction of a considerable part of the organisms inhabiting the planet. A global ecological crisis could arise not only as a result of changing temperature, but also by the organisms being affected by other atmospheric factors, in particular, by variations in the chemical composition of the atmosphere.

In studying the evolution of the atmosphere's composition it has been found that in the Phanerozoic, against the background of a general tendency to increase

6   1. Natural Climatic Catastrophes

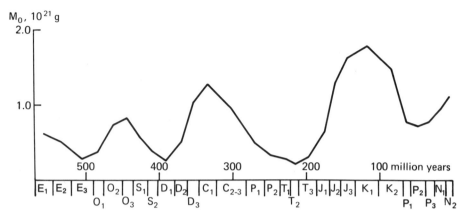

FIGURE 2. Changes in atmospheric oxygen mass (M_o) in the Phanerozoic.

the amount of oxygen in the atmosphere, several epochs occurred when the oxygen content decreased. As can be seen from Figure 2 (Budyko et al. 1985), the most noticeable reduction in oxygen occurred at the end of the Permian period and in the Triassic, when it was several times less than its maximum in the Carboniferous. It has been discovered that this decrease in oxygen mass coincided with the greatest ecological crisis in Phanerozoic animal history, in the course of which the variety of the majority of animal phyla considerably declined (Budyko 1982). It is likely that further studies will reveal other ecological crises (on a smaller scale, but nevertheless important) in the history of the biosphere.

Note that this section treats slow changes in the physical state and chemical composition of the atmosphere in the geological past. These changes could appreciably damage the biosphere and even destroy it under certain conditions. Possible effects on the biosphere of drastic climatic changes are discussed in Section 1.2.

PRESENT DAY CLIMATIC CHANGES

In comparison with climatic changes of the geological past, the present-day ones are much better studied thanks to the existence of a world system of meteorological observations over the last 100 years. It has become possible to calculate the annual mean temperature of the Northern Hemisphere by using data of surface air temperature obtained at meteorological stations. Similar temperature estimates derived from these data for the Southern Hemisphere were less accurate because of the lack of observational information. Since the general trends of mean temperature variations seem to be similar for both hemispheres, to study modern climatic changes we use the more reliable temperature data for the Northern Hemisphere only, where the network of meteorological stations is

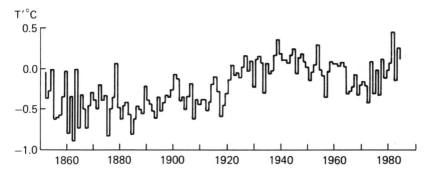

FIGURE 3. Anomalies of the mean Northern Hemisphere air temperature (T').

much better developed than in the Southern Hemisphere, rather than the data for the entire globe.

Figure 3 shows the values of air temperature averaged over the Northern Hemisphere. It is seen from the figure that throughout the last 100 years the trend to increasing temperature has prevailed. However, this trend was to a considerable extent disguised by comparatively short-term temperature variations. These variations do not disappear even on averaging the anomalies over 5-year and 10-year intervals (Kelly et al. 1985).

Although the global mean temperature variations for the last 100 years are relatively small, studying present-day temperature changes is of great importance for understanding the physical mechanism of drastically increasing climatic changes which can have catastrophic consequences.

The causes of modern climatic changes have basically been elucidated in the studies carried out over the past 20 years.

Although the idea of the effects of atmospheric transparency on climatic conditions was first proposed in the 18th century by Benjamin Franklin, it has not been widely accepted until comparatively recently.

The correctness of this idea has been proved by comparing variations in surface air temperature with those in solar radiation coming to the Earth's surface under clear sky and by using, for an explanation of solar radiation variations, calculations of radiation influx decrease due to scattering and absorption by stratospheric aerosol particles. Figure 4 gives the results of one of the preliminary comparisons between secular variations in mean temperature and direct radiation anomalies, the latter being averaged over large areas. Temperature and radiation anomaly estimates based on observational data are averaged for 10-year intervals. On comparing curves "a" and "b," a certain similarity is found between them. There are two maxima on both curves: one refers to the end of the 19th century and the other (major) to the 1930s. At the same time, there is some discrepancy between these curves: in particular, the first maximum is more noticeable in the secular trend of radiation compared to that of temperature. The similarity of curves "a" and "b" permitted the assumption that variations in radia-

1. Natural Climatic Catastrophes

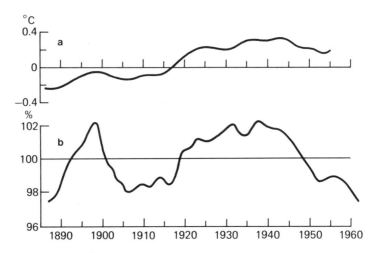

FIGURE 4. Anomalies of mean temperature *(a)* and direct radiation incoming to the Earth's surface under clear sky conditions *(b)*.

tion caused by the instability of atmospheric transparency are an important factor in climatic changes. To clarify this question, calculations have been performed of temperature variations due to fluctuations in atmospheric transparency for shortwave radiation which occur on increasing or decreasing the stratospheric aerosol mass. This analysis establishes that the calculated anomalies of the mean air temperature of the Northern Hemisphere agree well with the observational data from 1910-1960 (Budyko 1969). This means that basic features of climatic change for this period (including the warming of the 1920-1930s) can be attributed to atmospheric transparency fluctuations. Later, a similar conclusion was drawn by Oliver (1976), Pollack et al. (1975), Pollack et al. (1976), Karol (1977), Karol and Pivovarova (1978), Hansen et al. (1981), etc.

On studying modern climatic changes it has also been discovered that over the past several decades, along with atmospheric transparency fluctuations, the mean air temperature has been more affected by increasing carbon dioxide content of the atmosphere, described in more detail in Section 2.1.

Returning to the physical mechanism of the impact of the unstable content of stratospheric aerosol on climate, it should be mentioned that the chemical composition and the methods of formation of this aerosol were known back in the 1960s. Direct observations of stratospheric aerosol composition, initiated by Junge (1963), have shown that the aerosol consists basically of droplets of water solution of concentrated sulfuric acid. Most of the stratospheric aerosol particles are localized in the layer several kilometers thick, with the center being at heights of 18 to 20 km. This aerosol is formed by gases containing sulfates coming into the atmosphere from the Earth's depth, in particular during volcanic eruptions.

As can be seen from theoretical calculation and observation, if the source of stratospheric aerosol is localized in extratropical latitudes, aerosol spreads comparatively rapidly over the corresponding hemisphere and more slowly over the other hemisphere and, if close to the equator, it is distributed rapidly over both hemispheres. Particles of stratospheric aerosol gradually fall out due to both the force of gravity and large-scale air motions transferring them to the troposphere, from where they are quickly removed by precipitation. The average residence time of stratospheric aerosol particles (i.e., the time during which their number decreases by "e" times) is 1 or 2 years.

Solar radiation passing through the stratospheric aerosol layer dissipates partly due to absorption, but mainly due to its scattering on the aerosol particles. Since the typical size of most aerosol particles is of the order of tenths of micrometer, the scattering function is strongly stretched out in the direction of the falling ray. In this connection, direct radiation passing through the aerosol layer decreases much more strongly than the total solar radiation. A decrease in the full income of short-wave radiation coming to the troposphere noticeably affects the climate.

Since the particles of the indicated sizes affect the long-wave radiation less than the short-wave, the radiation balance of the Earth decreases. The stratospheric aerosol mass grows more noticeably after explosive volcanic eruptions, in the course of which a considerable amount of sulfur-containing gases enter the atmosphere, producing sulfuric acid droplets and silicate particles attenuating the influx of short-wave radiation to the troposphere. Climatic effects of volcanic eruptions are treated in more detail in Section 1.2.

1.2 Catastrophic Climatic Changes

LOCAL AND GLOBAL CLIMATIC CATASTROPHES

The term natural catastrophes is used in this book to signify abrupt changes in the environment that lead to the mass extinction of living organisms. It is clear from this definition that although the process mentioned in Section 1.1, of decreasing diversity of organisms with decreasing amount of oxygen at the end of the Paleozoic and the beginning of the Mesozoic, represented a global ecological crisis, it could not, however, be considered an ecological catastrophe, because it continued for tens of millions of years.

In comparison with global climatic catastrophes, local ones induced by atmospheric factors occur much more frequently. These local catastrophes, because of their limited duration, are associated in many cases with weather changes and not with climate variations. However, in the absence of a universally adopted time scale distinguishing the synoptic processes determining weather and the climate-forming processes, it is not always easy to discriminate between weather and climate catastrophes. We will consider several examples of frequently occurring local ecological catastrophes caused by atmospheric factors.

The best-known example of this kind is large-scale droughts covering areas of thousands of square kilometers. During such droughts, a considerable part of the natural vegetation cover and the crops perish. Most animals inhabiting the drought area also die or, if possible, move to other regions, which also results in the death of a part of forced emigrants.

The largest droughts in developing countries (pre-revolutionary Russia included) caused mass mortality in agricultural populations. Often epidemics spread far beyond the boundaries of the drought regions. The catastrophic droughts in the Sahelian region, Ethiopia, Sudan, and other countries are widely known recent examples.

Aftereffects of strong thermal variations disastrous for animate nature are also known and are frequently distributed over far greater areas than those affected by drought. These weather changes result less in mortality (although this always increases with excessive temperature variations), but inadvertently in mass deaths of animals and plants belonging to species sensitive to the thermal conditions of the environment.

Such an event is often observed in the mid-latitudes over vast areas. Long-term, strong reductions in temperature cause a considerable decrease in the number of many animal species, even full extinction. Particularly strong cooling reduces the area occupied by less frost-resistant plants.

It is less known that unusually hot weather can also cause mass deaths of animals. Thus, for example, a host of birds perished in Southern and Central Australia in 1932, when for more than 2 months the air temperature exceeded 38°C (Lack 1954).

One more example of comparatively frequent local climatic catastrophes is associated with the anomalous development of atmospheric processes when warm tropical waters move southward along the western coast of South America — the so-called El Niño phenomenon. In this case, the nutrient and oxygen-rich cold waters of the Peruvian current are overlapped over long distances by warm waters, which destroys the nutrient links supporting the existence of large fish populations. The disappearance of fish results in the mass deaths of birds feeding on fish, and creates the threat of starvation for the populations of coastal regions whose existence depends on fishery. Without dwelling on further aspects of the El Niño effect on atmospheric and ecological processes, we mention that this same effect can be observed at great distances from the region where this current is located (ref. Fedorov 1984 etc.).

The general feature of all local climatic catastrophes is that, although in a number of cases these catastrophes result in disastrous consequences, they rarely cause complete extinction of various species of animals and plants. First, this can be explained by the fact that most living organisms can renew their population even after great reductions and, second, that these organisms can survive in areas with more favorable environmental conditions. The exception to this rule may be less numerous organisms occupying small areas, i.e., that were on the verge of extinction before the climatic catastrophe.

Passing to global climatic catastrophes, these are associated with the extinction of many species of organisms as a result of drastic changes of climate either over the entire globe or over a part which is so great that it covers areas where a number of representatives of animal and plant kingdoms live. It should be particularly emphasized that the time factor is of great importance in the occurrence of global climatic catastrophes.

Even comparatively great climatic changes developing for a period of many thousands or even millions of years have not led, as paleontological data show, to mass extinction of organisms. The most striking examples of these climatic changes pertain to cases of vast glacier advances, in particular to the well-studied Pleistocene glaciations.

In glacial epochs of this time the climatic conditions drastically changed in those regions of middle and high latitudes where glaciations advanced. At the same time, the climate changed at all latitudes, including the tropics, where with some cooling the moisture conditions varied noticeably.

These considerable climatic changes, greatly affecting living nature, did not result in the mass extinction of organisms. The cause of this is very simple: with slow climatic changes ecological systems have not been ruined. These systems were preserved in certain geographical zones, which shifted as the climate changed. In middle latitudes of the Northern Hemisphere these zones usually shifted southward.

At the same time, gradual changes developed both in the structure of geographical zones and ecological systems, which favored the evolution of the organisms involved in these systems. However, in these cases there was no mass extinction of organisms.

It is easy to imagine what would have happened to animals and plants if such climatic changes had developed not over thousands of years but in the course of 1 or 2 years. Undoubtedly, in this case a massive ecological catastrophe associated with the extinction of a host of animal and plant species would have occurred.

Of great importance for clarifying the mechanism of global ecological catastrophes caused by increased aerosol mass in the stratosphere is the estimation of the effects of an abrupt drop in environmental temperature on animals and plants.

As noted above (Budyko 1971, 1984), an abrupt temperature decrease affects the life of different organisms in different ways. Limits in temperature decrease exist for every organism, which can: (a) result in its death; (b) weaken its activity so much that it will die in the course of the struggle for existence; (c) attenuate its resistance to infection, after which it inevitably falls a prey to infectious diseases; (d) disturb the reproduction process; (e) destroy plants and animals on whom the organisms depend (e.g., nutrition sources).

It is clear from this that actual values of a lethal temperature drop can be far lower than those that cause the death of organisms as a result of freezing. For ancient organisms these values are difficult to determine from modern ecological data, so that records of mass extinctions in the critical epochs of the geological past are of great importance.

The relatively short-term intervals during which the most considerable variations in successive floras and faunas took place are usually termed the critical epochs. The question of critical epochs has been discussed for many years, but even now there are still different opinions on this question, which is considered in more detail in Section 1.3. To understand this problem, we treat here the

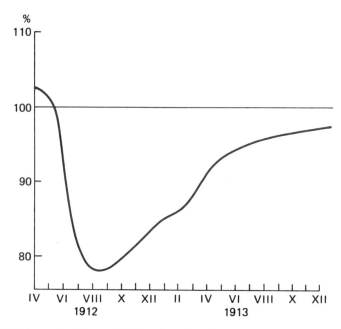

FIGURE 5. Changes in direct radiation (%) after the Katmai eruption (Alaska), spring 1912.

related question of the causes of climatic catastrophes that have occurred throughout the Earth's history.

VOLCANIC ERUPTIONS AND CLIMATE

As mentioned above, Benjamin Franklin was the first to pay attention to the possible climatic effects of volcanic gases and dust. He proposed that a large eruption of the Lacki volcano in Iceland in 1783 resulted in "dry fog," i.e., haze that caused a cold summer and poor harvests (ref. Humphreys 1940), in Europe. Later, Savinov (1913), Kimball (1918), Kalitin (1920) and other authors established that after an explosive volcanic eruption, solar radiation to the Earth's surface decreases drastically. In these cases, the value of direct radiation averaged over a large area can decrease by 10–20% over several months or years (Fig. 5). Figure 5 shows the ratios between monthly means of direct radiation under clear sky and after the eruption of the Katmai volcano in Alaska in spring 1912. This curve, based on observational data obtained at several radiation stations in Europe and America, demonstrates that in individual months the atmospheric aerosol attenuated direct radiation by more than 20%.

Figure 6 depicts the mean air temperature anomalies which would have arisen after this eruption in the absence of thermal inertia of the climatic system (curve 1). Actual yearly temperature variations decrease after the eruption by a factor

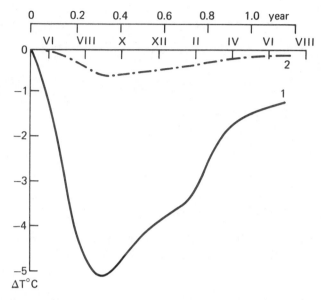

FIGURE 6. Changes in surface air temperature (ΔT) after a volcanic eruption.

of approximately 10 due to the considerable thermal inertia of oceanic waters. Temperature anomalies, calculated taking the thermal inertia of the climatic system into account, are depicted by curve 2. As can be seen, in this case decrease in temperature comprises only several tenths of a degree after the eruption, although the duration of the period of decrease increases somewhat compared with the inertialess climatic system.

Since the initiation of a world meteorological network (that started functioning approximately in 1880), several large eruptions of explosive type have occurred; one of them took place not long ago (in 1982, El Chichon volcano, Mexico). The eruption of Krakatoa in 1883 (Indonesia) was the largest explosive eruption that occurred during this period. After this eruption, approximately 20 km^3 of pumice and ash were introduced into the atmosphere. As a result of this explosion, oceanic waves formed, which crossed the Pacific and came to the Atlantic Ocean. After this eruption, unusually bright sunsets were observed all over the world, which are explained by the effects of the sharp increase in stratospheric aerosol mass (Rust 1982).

It is difficult to determine exactly the surface air temperature reduction after the Krakatoa eruption, since at the beginning of the 1880s the world meteorological network was very thinly spread and imperfect. The mean temperature decrease in the Northern Hemisphere can be assumed to have been approximately 0.5 °C.

A systematic study of the effects of individual volcanic eruptions on the Earth's surface temperature has recently been carried out by Kelly and Sear (1984). They used data from the largest eruptions over the last 100 years and by the so-called

14   1. Natural Climatic Catastrophes

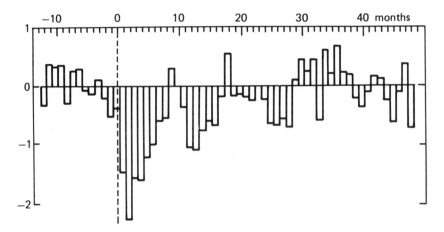

FIGURE 7. Mean changes in surface air temperature after volcanic eruptions: y axis = temperature anomalies in standard deviation units.

superimposed epoch method calculated, the yearly changes in monthly mean Northern Hemisphere temperature following every eruption, the results being averaged. Data obtained in these calculations are presented in Figure 7. They characterize mainly air temperature over land. As is seen from Figure 7, the greatest significant differences, reaching 0.3°C, are recorded during the second month after the event. This figure also shows that to reestablish the previous temperature, at least several months are required, which can be an estimate of the time of atmospheric response to similar influences.

The Mt. St. Helens eruption in the Northwestern United States on May 18, 1980 was well recorded. The volcano erupted a great ash train that was rapidly distributed over the eastern part of the state of Washington and over the nearby states of Idaho and Montana. Mass and Robock (1982) processed meteorological data for the territory over which the volcanic cloud passed. Having compared temperature measurement data over this territory and in the neighboring regions outside the cloud, these authors found that the day temperature in the areas under the cloud dropped by 8°C, probably due to drastic decrease in the level of solar radiation at the Earth's surface. However, the night temperatures in the areas under the cloud proved to be elevated by 4–6°C. Both effects can be attributed to the properties of the dust cloud containing particles with maximum sizes of 1 to 10 μm scattering sunlight strongly. According to estimates by Hobbs et al. (1982), the cloud contained about 2 Mt of particles with a diameter of more than 2 μm and an amount 200 times greater (by mass) of particles with a diameter of less than 2 μm. Particles of these sizes absorb (and emit) heat radiation well. At night, with heavy cloud cover, when diurnal temperature variations at the Earth's surface are weak, the ash cloud prevents the lower air layer from cooling down.

Observations of this phenomenon made it possible to determine the Earth's temperature change rate with decreasing sunlight intensity at the surface.

Far greater explosive eruptions than Krakatoa occurred in earlier times. During the last 200 years the eruption of the Tambora volcano (1815, Indonesia) was the largest, as a result of which 150–180 km^3 of pumice and ash were introduced into the atmosphere.

From the limited meteorological observations available at that time, it is difficult to estimate accurately the average air temperature reduction after the Tambora eruption. However, it is clear that this reduction was uneven and in a number of regions attained several degrees. In particular, in the summer of 1816, in Europe and North America the temperature was so low that the year was called "a year without summer" (the cause of this was unknown at that time). The eruption of Tambora deserves attention, because it is the eruption nearest in time that induced a climatic change which, in spite of its comparatively short duration, caused noticeable damage to living nature.

In particular, due to drastic decreases in crop yield in a number of regions far from the volcano, many thousands of people died of starvation (Stommel and Stommel 1979, 1983).

For the entire historical period (i.e., the past several millenia), data on large explosive eruptions can be obtained by chemical analysis of ice layers in vast glaciers formed over long time intervals.

These layers preserve sulfur compounds that settled on the ice surface from the atmosphere after explosive volcanic eruptions (Hammer et al. 1980, etc.). These data allowed us to establish that a great explosive eruption occurred in 536 A.D. Historical chronicles of late antiquity confirmed the formation of a slightly transparent haze in the atmosphere in that year that remained for more than a year. The brightness of the Sun was reduced by this haze to that of the Moon. Analysis of available data on this eruption led to the conclusion that it occurred in the tropics and that the aerosol cloud produced by the eruption was twice as dense as the aerosol layer resulting from the Tambora eruption.

Unfortunately, there is only limited evidence about the ecological aftereffects of this eruption. It is nevertheless known that in that year fruits did not ripen in the countries of the Mediterranean basin and Mesopotamia (Stothers 1984).

It is possible that the climatic change after the eruption of the volcano on the island of Santorin in the eastern part of the Mediterranean Sea, which took place approximately in 1500 B.C. was much more significant. Although the ecological aftereffects of this eruption are difficult to estimate, it is assumed that they were very considerable and led, in particular, to an abrupt decline of the highly developed Cretan civilization that had flourished up to that time. It is possible that this eruption is recorded in folk tales about the "outer darkness" recounted in the Bible (Rust 1982).

There is no doubt that for far longer time intervals, comparable with geological epochs and periods, the effects of volcanic eruptions on climate and the biosphere were much greater than the effects of the eruptions that have occurred over the last several thousand years.

It was known long ago that possible deviations from the norm of intensity indices of many natural processes increase as the time interval under consideration increases. Thus, for example, a catastrophic earthquake, whose probability is very low for a short time interval, is quite probable over a longer time interval.

It is natural to assume that massive anomalies in natural processes, that could not be observed over a comparatively short period of mankind's existence and for the short time of environmental study, had occurred throughout the long geological history of the Earth.

On the basis of this conception, the theory was formulated of the occurrence of aerosol climatic catastrophes in the past (Budyko 1969, 1971). This theory is based on the following ideas:

If the influence of individual explosive eruptions on the Earth's temperature is comparatively small, due to the limited amount of aerosol that the stratosphere gains after every eruption, it is obvious that the Earth's temperature changes much more drastically when many explosive eruptions occur one after another within a short time interval. The possibility of such coincidences for long periods of time, as a consequence of general statistical rules, increases noticeably with variations in the mean level of volcanic activity. Similar to this, the greatest amount of aerosol entering the stratosphere during one volcanic eruption will rise as the studied time interval grows, due to the same causes that induce the greater frequency of eruptions.

By analyzing empirical data on the influx of aerosol to the stratosphere over the last 100 years (Lamb 1969), the formula has been derived that relates the maximum aerosol amount entering the atmosphere within a certain time (t) depending on the total duration of the period (T), for which the corresponding analysis is carried out. It follows that the maximum income of aerosol with $T \gg t$ is proportional to the logarithm of ratio T/t.

Using this formula, it was deduced that for sufficiently long periods of time (thousands or millions of years), the amount of aerosol coming into the atmosphere during 10 years could exceed by 10-20 times the amount of aerosol contained in the stratosphere after the Krakatoa eruption.

The importance of this conclusion is illustrated in Figure 6. As seen from Figure 6, with a comparatively small explosive eruption the mean air temperature could decrease by almost 5°C with small thermal inertia of the climatic system. Since the thermal inertia of our planet is determined to a great extent by heat exchange in the mixed upper layer of the ocean, it is clear that, without oceans, the temperature drop after explosive eruptions would comprise not tenths of a degree but values of the order of 5°C. Such a cooling would undoubtedly have catastrophic aftereffects for living nature.

Simple calculations show that with the residence time of the aerosol layer in the stratosphere being 10 years, the influence of thermal inertia on temperature decline decreases considerably and becomes insignificant in the case of longer-term decrease of atmospheric transparency. Thus, a series of 10 explosive eruptions more or less evenly distributed over a 10-year time interval can induce a global climatic catastrophe. A similar catastrophe occurs after a single explosive eruption, when an aerosol layer is formed in the stratosphere, that is denser by one order of magnitude than the aerosol layers that appeared after the Katmai and Krakatoa eruptions.

It is possible to estimate the total mass of particles composing this aerosol layer. Even at the beginning of the 1970s, from data obtained in several empirical and theoretical studies, the mean value of parameter M was found equal to the mass of optically active particles in the vertical column of atmospheric air attenuating the flux of total radiation by 1%. Parameter M was found to be approximately 10^{-6} g cm^{-2} (Budyko 1974). Taking into account that a 1% decrease in radiation entering the troposphere reduces the mean surface temperature by 1.5°C with a steady-state climatic system, and that the thermal inertia effects decrease this value by one order of magnitude, we find that with the mass of optically active aerosol particles equal approximately to 10^6 tons, the mean surface air temperature is lowered by 3°C. The total aerosol mass in this case somewhat exceeds the indicated value.

Note that the stratospheric aerosol particle mass derived from data of direct observations after the recent explosive eruption of the El Chichon volcano comprised several million tons (Rampino and Self 1982). It is clear from the above estimate that with this amount of aerosol the mean air temperature could decrease by about 0.1°C.

In a recent study by Groisman, it was discovered that temperature actually decreased a little after the El Chichon eruption. However, an accurate value of this decrease appeared to be rather difficult to determine, since during several years preceding the eruption the mean temperature for individual years had varied within wide limits. Probably this problem will be easier to solve some years after the eruption, by using temperature data pertaining to the period before and after its occurrence.

The above conclusions about the physical nature of aerosol climatic catastrophes occurring after explosive volcanic eruptions were presented in studies published at the end of the 1960s and the beginning of the 1970s. Of later studies on this problem we mention the calculation carried out by one of the present authors together with Groisman. In this calculation, to estimate the probability of aerosol catastrophes induced by volcanic eruptions in the geological past, methods of mathematical statistics were applied, taking into account data obtained in geochemical studies of variations in volcanic activity level during different epochs of the Phanerozoic. The analysis confirmed the conclusion as to the possibility of the formation of aerosol layers over sufficiently long time intervals, with the mass inducing the reduction of global mean surface air temperature by 5–10°C or more. Since in this case air temperature on the continents decreased by far greater values than for the planet as a whole, such coolings could have led to the extinction of numerous species of animals and plants.

The above simulated estimates of change in air temperature as a result of aerosol layer density are based on the assumption that temperature changes are directly proportional to a decrease or increase of the initial aerosol layer density.

This assumption is undoubtedly approximate, since the density of the aerosol layer varies with time, the rate of decrease in density that determines the lifetime of aerosol particles increasing due to the intensification of coagulation and sedimentation processes by increasing the initial density. At the same time, with the high density of the aerosol layer, it is necessary to take into account the nonlinear relationship between the attenuation of short-wave radiation flux and aerosol layer density. It is possible, however, to show that this nonlinearity is not very important for approximate estimates of an insignificant increase in aerosol layer density.

This question is treated in studies on the formation of the stratospheric aerosol layer due to natural and anthropogenic factors (Asaturov 1977, 1979, 1981, 1984a,b). A model for calculating the transformation of aerosol particles not containing sulfur (for example, dust and smoke particles), injected into the stratosphere for the homogeneous aerosol layer has been developed by Asaturov (1977, 1979). This model takes into account laws governing coagulation, sedimentation, and vertical turbulent transfer of aerosol particles.

Later, together with these processes, the model took laws governing the income to the stratosphere of sulfur-containing gases into account and their reduction to sulfuric acid vapor as well as the absorption of this vapor by stratospheric sulfuric acid aerosol (Asaturov 1984a). Consideration of these processes is important in simulating background and volcanogenic stratospheric aerosol containing sulfur.

In contrast to the model of Turco et al. (1979), that developed by Asaturov makes allowance for the nucleation of sulfuric acid vapor and the tropospheric sink of formed nuclei. With these processes accounted for, the predicted increase in optical density of background stratospheric aerosol layer with increasing ejections into the troposphere of sulfur-containing gases is twice the value obtained without considering the indicated processes.

The question of the relationship between the initial density of the aerosol layer and its influence on the radiation regime is considered in Asaturov (1984), who applies a theoretical model of stratospheric aerosol layer evolution to calculating its density changes with time. In this study the effects on the evolution of the stratospheric aerosol layer of changes in its initial density were determined over wide limits up to values exceeding these characteristic of modern explosive eruptions by 10,000 times. Restricting ourselves to the maximum values of initial density important for our theme exceeding the indicated ones 100 times, we note that with a 100-fold increase in the initial density the characteristic time of aerosol particle life decreases, due to coagulation, by approximately two times compared with the value typical of modern explosive eruptions. With a 10-fold increase in the initial layer density, the radiative energy flux over the time of the existence of the layer usually attenuates not by 10 but approximately by five times. These deviations in the relationship between the anomalies of heat radiation influx and variations in aerosol layer density from direct proportionality are not very significant for preliminary studies concerning the problem of aerosol catastrophes. However, they should be taken into account in a more detailed analysis of the problem.

As noted in this study, because of the nonlinearity of the dependence of the Earth's thermal regime on the mass of ejection, the temporal and spatial structures of ejection into the stratosphere are essential for the formation of climate. For instance, a series of ejections approximately similar in power can lead to noticeably greater lowering of the average surface air temperature than one more powerful ejection into the stratosphere, whose mass is equal to the sum of the series under consideration. If a number of ejections occur at the same time into the stratosphere, the lowering of the mean surface temperature will be greater, the better the homogeneity of their distribution over the globe. Both these effects are stronger, the greater the total mass of ejections into the stratosphere.

The Report of the US National Research Council (1985) discusses the possible climatic role of the largest explosive eruptions (see also Francis 1983). As the authors of the Report believe, large eruptions with a volume of aerosol particles ejection greater by one order of magnitude than the ejection during the Tambora eruption can produce an aerosol mass of about 10^9 tons. According to the rough estimate given, such a quantity of dust and sulfates in the stratosphere would cause an average temperature drop, for the Earth as a whole, of 10°C over several months. Note that this calculation is rather poorly grounded. In particular, the absolute value of temperature reduction obtained after an eruption is probably underestimated.

As indicated further, with an aerosol catastrophe the mean temperature drop on the continents should be several times greater than the average global reduction. It is absolutely certain that an abrupt cooling by tens of degrees would lead to the death of the majority of animals and plants inhabiting the land. Along with decreasing intensity of photosynthesis, comparatively small (of the order of several degrees) lowering of the sea water temperature can annihilate stenotherm forms of marine organisms.

The Falling of Celestial Bodies and Climate

The hypothesis of the possibility of aerosol climatic catastrophes as a result of the collision of celestial bodies with the Earth was proposed at the end of the 1970s. Here we shall briefly explain this hypothesis (Budyko 1980). It is shown in this monograph that one of the results of the falling of large meteorites should be a considerable increase in aerosol layer optical density in the atmosphere. Preserved traces of meteor craters on the Earth's surface allow us to assume that during the long history of the Earth, meteorites, of a size up to several hundred meters or even more, collided with the Earth. If, as a result of the explosion caused by a meteorite falling on the Earth's surface, the stratosphere gained aerosol particles of even only a small portion of its mass, it would be sufficient to cause an abrupt decrease in the solar radiation reaching the Earth's surface.

By calculation it was concluded (Budyko 1980) that after the falling of a sufficiently large meteorite the temperature would be lowered by approximately 5-10°C for many months, and this would catastrophically affect various living organisms. The question of effects in the biosphere caused by the collision of celestial bodies with the Earth attracted great attention after the appearance of publications by Alvarez et al (1980, 1982), which are discussed in Section 1.3.

A detailed calculation of climatic effects of the Earth's collision with a large asteroid was carried out by Toon and coauthors (Toon et al. 1982; Pollack et al.

1983). In these studies it was concluded that, due to coagulation and sedimentation of the aerosol particles produced by the explosion of the asteroid, the concentration of these particles remains high only for several months after the asteroid falls. At this time, the solar radiation influx decreases to a level insufficient to support photosynthesis. This calculation shows that attenuating the solar radiation influx would reduce the mean surface air temperature over the ocean surface by 2 or 3 °C for 2 or more years, and over the continents by several tens of degrees for half a year. It was also determined in these calculations that for 10 months after the falling of an asteroid, the global mean temperature is reduced by 9 °C and in 20 months, by 6 °C, on the average. These results are close to values obtained in the calculation by Budyko (1980).

On analyzing annual variations of temperature, it can be concluded that Toon et al. (1982) overestimated the reduction of the difference between ocean and continental temperatures due to attenuating radiation. Probably the authors did not account for the temperature effects of increasing ocean-continental heat exchange in the atmosphere after the fall of the asteroid, which should noticeably decrease the difference between temperature reductions over land compared with those over the ocean.

Shoemaker (1983) found that if for bodies of radius r in kilometer size range the frequency of collisions is proportional to r^{-2}, then with $2r = 10$ km, this frequency corresponds approximately to one collision per 10^8 years. With the characteristic rate of collision of 20 km s^{-1}, this body striking against the Earth releases the energy amount of about 4×10^{23} Joules [or 10^8 Mt in trinitrotoluol (TNT) equivalent]. The thickness of the sedimentary layer dividing the Cretaceous and Tertiary periods comprises, on the average, about 2 cm (where it can be discerned). With this thickness of sedimentary layer, the total mass of global sediments can be estimated at 2×10^{13} t. It is interesting that a recent study, including electron microscope analysis of clay particles in this layer, showed that they are fused or have the appearance of particles that had been affected by high pressure. This is additional evidence of the impact origin of the layer (Kastner et al. 1984).

It has been assumed that some impact craters fit this event in age. In a study by Masaitis and Mashchak (1982), attention was paid to the Kara crater located on the river Kara, rising in the North Urals and falling into the Baidaratskaya inlet of the Kara Sea. According to recent data (V.I. Feldman, pers. commun. 1984), its age is 63 ± 3 million years. The Kara crater, or Kara depression, in Masaitis' terminology, represents an impact crater: actually it consists of two craters, the main, Kara, of 60 km in diameter, and the other, Ust' Kara, of 25 km in diameter, stretching out in part to the bottom of the Baidaratskaya inlet. Of the same age is the Kamensky crater located near the town of Rostov on the Don with diameter of 11.5 km; all of them are on the same arc of a great circle and could have been formed as a result of the falling of two or three bodies which were bound in space by weak gravity into a single system and moved along the same trajectory, disaggregating in the Earth's strong gravity field as they approached the planet. Such a system, after hitting the Earth, could raise an amount of dust into the air sufficient, by order of magnitude, to produce the described catastrophe, since the simultaneous collision with the Earth of two (or three) asteroids releases more energy and forms more dust than one asteroid with mass equal to the sum of these two (or three) asteroids.

We do not know the initial size distribution of particles formed by impact collisions. However, following the analogy with a nuclear burst, when 1 Mt of explosive power produces approximately 0.3 Mt of dust by mass, then we obtain 3×10^{13} t of dust. This agrees well with the estimate of 2×10^{13} t derived from the global thickness of sedimentary layer of 2 cm given above.

The fact that the released energy exceeds by at least four orders of magnitude the explosion energy of a nuclear bomb in the event of a nuclear war makes the possibility of climatic catastrophes after such collisions quite clear. It is obvious that the collision with a body of 5 km in diameter will produce correspondingly less dust; however, even this could be sufficient to fill the global atmosphere with dust.

A 0.2-cm sedimentary layer formed in this case would have been more difficult to retain in the geological records of the Earth; however, a concentrated search for geological and biological evidence of such collisions should be undertaken. Toon and his co-authors calculated that the ejection of 10^{11} t of dust (the amount to be expected from the collision of a body with a diameter of 2 km) would lower the level of solar radiation flux to such an extent that it would be insufficient to support photosynthesis. This would result in almost the same sharp surface air temperature drop as if 10^{13} t of dust were ejected. This can be attributed to the strong nonlinear dependence of temperature reduction on the level of illumination. This is discussed in Section 2.2 (Toon et al. 1982; Pollack et al. 1983).

The dependence of solar radiation influx on the amount of dust in the atmosphere is shown in Figure 8 (Toon et al. 1982; ref. also Pollack et al. 1983). The vertical axis in this figure represents the fraction of sunlight reaching the Earth's surface and the horizontal axis, the optical density τ. For dust (with an average radius of 0.25 μm logarithmically normalized particle size distribution at variance parameter 2 and complex refraction index $m = 1.5-0.001i$) the coefficient of light extinction with the characteristic wavelength of solar spectrum 0.5 μm is equal to 2.8 m² g⁻¹. If all the particles in the dust layer of 2-cm-thick and 2-g cm⁻³ density were of sub-micron size, the optical density, τ, would be estimated as $\tau \cong 10^5$. Figure 8 shows that the values of τ, even of the order of 100 for dust, decrease the level of illumination by seven orders of magnitude, i.e., draw nearer to the limit of the sensitivity of the human eye.

The information given in this section indicates that in the geological past, large short-term climatic changes occurred, which could have affected living nature.

At present, only one physical mechanism of these changes is known – a considerable increase in atmospheric aerosol mass which resulted in decreasing atmospheric transparency and could have caused aerosol climatic catastrophes.

The possibility of these catastrophes occurring follows from the fact that a reduction of mean air temperature after the large explosive eruptions of the past centuries (Krakatoa and Tambora) was less in order of magnitude than would

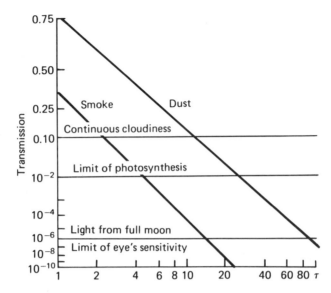

FIGURE 8. The dependence of sunlight transmission by dust and smoke layers on their optical depth (τ).

have resulted in a massive aerosol climatic catastrophe. Taking into account that volcanic activity changes considerably with time, and considering in terms of statistics the above possibility of the coincidence of a series of eruptions over short intervals, it is clear that throughout the Earth's history an increase in the density of the volcanic aerosol layer, by at least one order of magnitude (and possibly by more) could have taken place many times.

It is also beyond doubt that a considerable increase in the density of the aerosol layer occurred after the fall of large celestial bodies on the Earth's surface, although it is much more difficult to estimate a decrease of atmospheric transparency after these events. Additional information about this second manner of formation of aerosol catastrophes is given in Section 1.3.

To understand the physical mechanism of aerosol climatic catastrophes, data on dust storms arising on our planet and, on far larger scales, on Mars are important.

Dust Storms

Dust storms can cause considerable changes in local meteorological conditions on the Earth. Dust particles form clouds of 2 or more km thick. Dust can effectively attenuate solar radiation, mainly through scattering and partly by absorption. As a result, the atmosphere is heated, which reduces the temperature decrease with height. This increases the atmospheric stability. There exists another, purely hydrodynamic cause of increase in air flow stability and attenuating the turbulence in it, the theory of which was developed for these conditions by Barenblatt and Golitsyn (1974). They proposed a turbulent boundary layer model for the flux containing a heavy admixture with different levels of temperature stratification. In this flux, air-borne dust can be suspended due to velocity-turbulent fluctuations, i.e., a portion of the turbulent kinetic energy should be expended to support the suspension of dust. As a result, other things being equal, the turbulence intensity in the dust flow is lower than in a clear flow. Therefore, in the dust flow, the turbulent mixing is suppressed. This suppression increases as the amount of the admixture grows. This can lead to attenuating the convection, to increasing wind velocity with height, even at several meters above the surface; the latter phenomenon observed in nature (see Barenblatt and Golitsyn 1974), could raise additional dust from the rough surfaces.

The same results are expected from the heating of dust clouds by solar radiation. In heated atmospheric areas the pressure decreases compared with the surrounding clear atmosphere. Winds appearing due to this fact can raise new amounts of dust (Golitsyn 1973b). Two indicated mechanisms (1) increasing wind in a flow with heavy suspension as a result of attenuating turbulence, and (2) wind systems arising due to inhomogeneous heating of the atmosphere by dust absorbing sunlight, constitute the so-called positive feedbacks that help the development of dust storms. They cease to develop further, when the size of the dust cloud exceeds that of the surface areas that could produce dust, or the wind system drives the cloud from these areas.

The largest dust clouds on the Earth are observed over the Sahara. Eastern and northeastern trade winds carry these clouds over the countries of Western Africa and bring them to the Atlantic and then to Florida and even Mexico. Early in the Middle Ages the part of the Atlantic near the western coast of Africa in the vicinity of the Cap Verde islands was called the "Sea of Darkness." Satellite observation data (Carlson 1979) showed that the optical density τ, equal to unity or more, caused by dust clouds, can be detected over an area of 10^6 km^2, and that such a cloud contains about 8 Mt of dust. Oceanographers often observe that the sea surface temperature below such clouds is lower than in adjacent clear areas.

In Western Africa, during the dry season (January–May) northeastern winds prevail, and often bring dust. This wind is called Kharmathan. Brinkman and McGregor (1983) indicate in their study that when such winds occur in Nigeria, the optical depth can reach 2. Total solar radiation (direct and scattered) in this case decreases by up to 30% and daytime temperature is lowered by 5–6 °C.

Far stronger are dust storms on Mars, where they frequently attain global scale, covering the planet with a thick shroud. Thus, in autumn 1971, when approaching Mars, the TV sets on the American automatic station Mariner-9 and Soviet stations Mars-2 and 3 registered that the whole planet was covered with a turbid shroud, above which only four cone-shaped peaks emerged. Later, the height of these mountains was determined as 15 km or more (see Moroz 1978). There are no oceans on Mars and a large portion of the planet appears to be covered with dust. Global dust storms are observed almost every Martian year. They usually start when Mars is located close to its perihelion. The ellipsoid orbit of Mars is so elongated that when approaching the perihelion, the planet receives from the Sun an amount of energy 1.5 times greater than it does in the aphelion. At this time in the southern hemisphere of Mars, the end of spring or the beginning of summer occurs. Dust storms usually begin in specific places, namely, in southern subtropical and temperate latitudes. Within a few days dust covers this latitude entirely and in approximately one week after this it spreads to the poles.

Data from planet indirect sounding (see Moroz 1978) indicate that Martian dust absorbs sunlight somewhat more strongly than Earth's silicious dust. When the intensity of the dust storm is maximal, the value of τ reaches 5. The thickness of the atmospheric dust layer can amount to 10 or even more kilometers. In this case, as direct measurements on the planet's surface show, the Martian atmosphere is heated by several tens of degrees (Moroz 1978), and the surface temperature decreases by 10–15°C (Ryan and Henry 1979). These measurements show that in the transparent Martian atmosphere, a regular cyclonic system is observed. However, when the dust cloud comes, any cyclonic activity disappears while this cloud exists, i.e., for a month or even more (Sharman and Ryan 1980). This demonstrates that in the atmosphere, made turbid by absorbing aerosol, the whole circulation changes. The formation of thermal and dynamic variations in a planet's atmosphere and their interrelations will be discussed in Section 2.2, where the Martian atmosphere will be treated in more detail.

In addition, information will be presented about Arctic haze—the phenomenon similar (in its effects) to dust storms. Recent studies have shown that in spring the Arctic atmosphere contains a great amount of aerosol (Patterson et al. 1982; Rosen and Novakov 1983). These are fine aerosol particles with modal radius of about 0.4 μm, rich in soot. They exist in the layers at a height of up to 5 km. The optical density of the Arctic haze reaches 0.3. Calculations show that such a layer can noticeably affect the albedo. However, this effect depends also on the Sun's zenith angle, i.e., on the geometry of illumination of the layer and on the albedo of the Earth's surface: over the open ocean with low albedo, the total local albedo can increase by several percent and over ice with high reflection, the total albedo can decrease by 10%. All this can result in local temperature changes within several degrees.

Arctic haze is also interesting by the fact that, due to the specific properties of the Arctic meteorological conditions, its residence time is several months. The polar front formed at the end of winter or in spring leads to the arising of a stable air mass in high latitudes, where precipitation is deficient. Therefore the processes of washing out the aerosol are strongly weakened and the aerosol evolution, as well as its optical properties, can be observed during several months.

Smoke from Forest Fires

This is the last example of natural events resulting in the increase of atmospheric aerosol—large forest fires. In the territory of the USSR, large forest fires have occurred frequently. The Nikon chronicle describes fires in the dry summer of 1371, when smoky haze covered vast areas. The Sun was dim and spots were seen on it "like nails," animals left forests and in the autumn the crop did not ripen, having yielded "meager grain." Large forest fires occurred in the summer of 1915 in Western Siberia and in the summer of 1972 in the European USSR, when fires occurred in forests and peat bogs, and smoky haze persisted for several weeks.

Smoke from large forest fires usually rises as high as 2–3 km. In August, 1972, smoke rose as high as 5 km. Satellite observations allow us to follow the evolution and trajectories of smoke clouds. Grigoriev and Lipatov (1978) describe how in August, 1972, a cloud of smoke formed in the center of the European USSR, rounded the Urals, and as a flow several hundred kilometers wide, reached Lake Balkhash.

The height of elevation of the smoke certainly depended on meteorological conditions, primarily the vertical gradients of temperature and wind. The greater they were, the lower the elevation of smoke and the quicker its dispersion. Large fires almost always occur under conditions of dry anticyclonic weather. However, in other cases, smoke can rise to even greater heights. Thus, in September 1954, when large fires occurred in Buryatia, a great amount of smoke was observed at a height of about 8 km (Korovchenko 1958). Some cases are known of smoke particles being detected in the stratosphere (Cadle 1972).

The most comprehensive description in literature is available about the evolution of a smoke cloud from forest fires that occurred in western Canada in the last decade of September 1950 (Wexler 1950; Smith 1950; Watson 1952). Fires covered an area of about 40,000 km^2 in the northwestern part of the province of Alberta and northeastern part of the province of British Columbia. Within 2 days after the beginning of the most intense phase of the fires, a smoke cloud was formed over entire Canada. It covered all the eastern states of the U.S., east of the Mississippi, and some other states (Fig. 9). In many cities at noon the illumination decreased so that streets had to be lit. Within 5 days the cloud had reached Western Europe and covered the area from Spain to Scandinavia. In England, where the cloud induced many unusual optical phenomena (blue sun and moon), special aircraft flights were organized, during which smoke particles were discovered at heights of 10 to 12 km (Bull 1951).

Detailed observations were carried out near Washington, D.C., the capital of the U.S. (Smith 1950). It was discovered that the smoke layer was located as high as 2.5-5 km and was sandwiched by inversions, whereas inside the layer the temperature gradient was adiabatic, which showed a good mixing of the layers. The total solar radiation was reduced by half (Wexler 1950). According to Wexler, all this was accompanied by a 4°C drop in temperature for 4 days (compared with the forecast for transparent atmosphere). According to Smith (1950), the maximum temperature drop comprised 6°C; however, at night the temperature did not rise, as was the case in the state of Dakota 30 years later (in 1980) as a result of the eruption of the volcano Mt. St. Helens (see above). This difference can be attributed to the fact that smoke particles that are much finer than volcanic ash absorb little of the thermal radiation of the Earth's surface and of the lower atmospheric layers, but increase scattering and absorbing solar radiation, thus barring its influx, to the surface of the Earth.

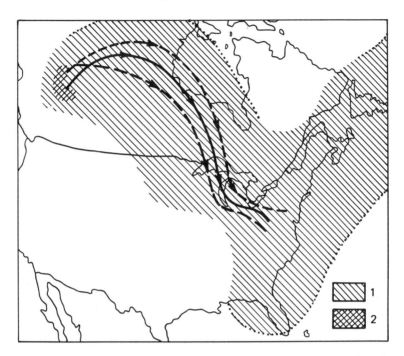

FIGURE 9. The spread of smoke cloud *(1)* from a fire source in West Canada *(2)* at the end of September 1950 (Smith 1950).

The study of large forest fires has established how high and how far this smoke can spread and how it influences the weather. The results obtained allow us to study the major physical properties of smoke, the laws governing its behavior in the atmosphere, and its role in changing meteorological conditions. At the same time, analysis of the effects of different natural and anthropogenic aerosols enables us to see the role of various properties of aerosols and their influence on different atmospheric processes, weather, and climate.

Studying the influence of variations in atmospheric aerosol mass on radiation and temperature in the atmosphere is of great importance in investigating the physical mechanism of the aerosol climatic catastrophes of the past which have been discussed above.

1.3 Critical Epochs of Geological History

MASS EXTINCTIONS OF ORGANISMS

Proposals that in the past global climatic catastrophes took place from time to time that greatly affected animate nature and led to the extinction of numerous organisms were made at the beginning of the 19th century by the founder of paleontology and comparative anatomy, Cuvier. Cuvier's hypothesis was probably based on the ideas of Laplace, who assumed that these catastrophes were caused by the falling of comets on to the Earth, which resulted in destructive earthquakes and the flooding of vast areas by tidal waves.

The concept of catastrophism exerted a noticeable influence on the origin of biostratigraphy, since it explained the abrupt differences in animate nature between certain geologic periods. It also explained the systematic changes of successive faunas reflected in the forms of widely distributed fossils.

By the mid-19th century, the concept of catastrophism was rejected by the majority of researchers mainly due to the views of Layell, who proposed an actualistic approach to studying the Earth's past. According to this view, these processes have operated, throughout the history of our planet, and are still determining its evolution at present.

Later, it became clear that actualism, that had so profound influence on the development of geology, is not a fully universal concept. In particular, Yanshin indicated that the processes determining the formation of past natural conditions differed in many respects from the modern ones. In this connection, the purely actualistic approach to studying the Earth's history frequently proves not to be correct (Yanshin 1961).

Throughout the past decades, prominent biologists have proposed many times that in the geological past drastic changes in abiotic conditions of the environment repeatedly resulted in the mass extinction of organisms (Shmalgauzen 1940; Mayr 1963). At the same time, individual scientists (particularly Davitashvili 1960) assumed that it was impossible to prove, given the incomplete paleontological records and unavoidable incorrectness of dating certain fossils, that the extinction of many species of organisms occurred simultaneously in the critical epochs of geological history. In this connection, they denied mass extinctions caused by abiotic factors.

The fact that mass extinctions occurred throughout the Earth's history (mainly at the end of geological periods) has been recognized by many researchers (see reviews by Newell 1967; Valentine 1968; Herman 1981, etc.), although this concept is not universally accepted.

It might be thought that in a number of cases the belief in the absence of mass extinctions results from an incorrect interpretation of incomplete paleontological records. In solving this problem, one should keep in mind that the conclusion about the existence of a given species (or genus) at any time is usually based on the relevant remains of animals available. If these finds are not available for any particular time interval or later, the conclusion is drawn that the given species (or genus) is extinct.

Both these conclusions are inaccurate. The first refers not to one certain moment, but to a comparatively long time period whose duration can depend on errors in dating the available find (or groups of finds). The second conclusion cannot be correct, for many different causes can prevent the preservation of the remnants of animals that once existed, the simplest being the decreasing number of animals and the area they occupy, after which the probability of preserving their remains is very slight. Thus, for example, it is very unlikely that in millions of years from now it will be possible, from paleontological records of our time to prove the existence of representatives of groups of animals that lived in the past, such as rhynchocephalians or crossopterygians. The reason given can be very important for the correct usage of paleontologic data in studying mass extinctions.

As noted above, some authors have doubted the reliability of data on mass extinctions, assuming that based on the limited accuracy of dating, the last finds of certain species of animals for various areas do not show that these animals were simultaneously extinct. It has also been mentioned that sometimes before an epoch of mass extinction the representatives of those groups of animals which finally disappeared completely during critical epochs had begun to be gradually extinct. The former idea of the limited possibilities of using paleontological data offers no solution to the question of short-term mass extinctions. The latter idea, while in some cases reflecting actual changes in fauna compositions, in others is simply a statistical result not pertaining to the extinction. If, because of the limited number of finds of animal remains, these are not available for every time interval studied, then unavoidably, before the actual mass extinction, the disappearance of this or that animal species (or genera) would be recorded, while they were still far from being extinct. The more limited the completeness of paleontological data, the more such "pseudoextinctions" appear to have taken place before the real disappearance of a relevant species or genera of animals.

This is a simple example to illustrate this phenomenon (Budyko 1982). Let us assume that during a critical epoch a certain number of animal groups died out practically simultaneously and that the probability of discovering remains of every animal group before the critical epoch for time interval T is equal to 50%. Then it is most probable that 50% of these animal groups would have disappeared in paleontological history not during the "critical epoch" but T years earlier, 25% would vanish 2T years earlier, etc. Since time T can comprise millions of years, it is easy to draw an erroneous conclusion that the process of extinction lasted for a long time and developed gradually, terminating in the critical epoch. Such errors are frequently not taken into account when studying the process of animal extinction.

All this refers to the organisms whose discovered remains are few compared with the number of their species and the duration of their existence, but dated validly; for example, the dinosaurs. This group existed for more than a hundred million years and included many species and genera. However, the remains of several thousand animals only have been found, which is insufficient for any reliable determination of the duration of the existence of every discovered genus or species. It is somewhat easier to study by paleontological data the processes of extinction of an organism for which data are available, such as for many marine invertebrates. However, even in this case there are numerous difficulties associated with the unavoidable inaccuracy of dating certain sediments. While mentioning the great success of modern stratigraphy in solving problems of rock age determination by isotope analysis (mainly by the K/Ar method), one should remember that these methods yield satisfactory results only when favorable conditions are available.

One well-known example demonstrates the possible errors that arise in determining absolute age. The finding by R.E. Leakey of the skull of *Homo habilis*, known as the skull N 1470, was considered to be one of the greatest discoveries of modern paleoanthropology. The importance of this find lay in its antiquity, as the age of the skull determined by the K/Ar dating method was 2.9 million years. This time interval was longer than the age of all other finds of remains of the representatives of the genus *Homo*, so that the find of Leakey was greeted as a scientific sensation.

Since the dating of the skull N 1470 attracted great attention, it was studied by using all modern methods of stratigraphy. As a result, in the opinion of most researchers, the primary age of the skull has been reduced by approximately 1 ma. This was done several years after the first determination of its age; the error of the first estimation of absolute age in this case amounted to about ⅓ of the value obtained. With this example in mind it is very difficult to believe that the relative error of determining the absolute age by the K/Ar method does not exceed 1%.

The answer to the question about the reality of mass extinctions can probably be obtained by using mathematical statistic methods. As a result of one such analysis, the probability of random coincidence in time of the extinctions of a number of reptile groups at the end of the Cretaceous has been calculated, taking into account possible differences of time of extinction of individual groups (Budyko 1971). This calculation showed that the probability of random coincidence of these extinctions is equal to 10^{-4}, i.e., it is negligible. Later, more detailed calculations were carried out by Raup (1979) and other authors, which in all cases confirmed the reality of mass extinctions during the critical epochs of geological history.

Let us discuss the results of one of these studies. The conventional rate for the extinction of the families of marine invertebrates and vertebrates, which changed comparatively little with time, was derived from data on variations of the number of these families. In comparing this rate with actual data on extinctions, it was established that at the end of five geological periods the extinction rate greatly exceeded the conventional one, and that in four cases (the late Ordovician, Permian, Triassic, and Cretaceous) the probability of the random coincidence of numerous extinctions was negligibly small. The fact of several mass extinctions in the Earth's history has thus been statistically confirmed (Raup and Sepkosky 1982).

Figure 10 presents Raup and Sepkosky's data on variations in the number of families of animal groups under consideration throughout the Phanerozoic. It can be distinctly seen in this figure that the number of families noticeably decreased during the epochs of mass extinctions. However, it is impossible to estimate directly from the data in this figure how many species of animals disappeared in the course of each extinction.

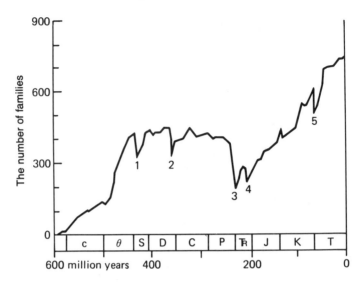

FIGURE 10. Changes in marine animal families during the Phanerozoic. *1–5* are the epochs of mass extinctions.

It is easy to understand that the relative number of genera which ceased to exist after a mass extinction should be greater than the relative number of families that disappeared and that the relative number of extinct species is greater than the number of genera. The cause of this is that in a number of families not all the genera become extinct and in many, not all related species.

Since reliable data can be derived from paleontological evidence only concerning the number of extinct higher taxonomic groups (for example, families), to determine the number of genera and species that disappeared, statistical estimates were used. By this method it was found that at the end of the Cretaceous about 50% of the genera that then existed and about 75% of the animal species disappeared (Russell 1979, 1982). An even greater extinction seems to have occurred at the end of the Permian, when according to available estimates, about 96% of marine animal species vanished (Raup 1979). Such scales of extinctions indicate that at certain moments of the biosphere's history the threat of destruction became more acute.

These results confirmed completely the views of the greatest biologists of our time, Schmalgauzen and Mayr, who even before these studies were completed, did not doubt the fact of mass extinctions having taken place in the critical epochs of the geological past due to the influence of abiotic factors. Another prominent biologist of the 20th century, G.G. Simpson said: "I agree that mass extinctions have occurred at certain times in Phanerozoic history and that they are distinct from the background extinctions that were happening in all geological epochs" (1983).

The Duration of Mass Extinctions

If the fact of mass extinctions in the critical epochs might be considered proven, the problem of determining the duration of such a period of mass extinction is less clear.

In the above-mentioned studies of Russell and Raup, it is noted that although paleontological data indicate that mass extinctions took place in the past, they do not answer the question of their duration. For example, Russell believes that these extinctions could have taken place both instantly and over ten or a hundred thousand years.

Raup and Sepkosky show by statistical analysis that the epochs of extinctions were comparatively short, but it is impossible to determine their duration by this method.

Attempts to estimate the duration of mass extinctions by direct analysis of paleontological data have led to controversial results even in the most recent studies. Many of these studies refer to the last of the most significant mass extinctions, which occurred at the end of the Cretaceous. At this time many representatives of marine and land flora and fauna vanished. At the end of the Cretaceous a considerable, perhaps the greater, portion of marine plankton species died out. Three of four families and 17 of 23 genera of planktonic foraminifera were extinct by the late Cretaceous. At this same time, most of the families and genera of bivalve mollusks disappeared, and many brachiopodes, ostracodes, ammonoidea, belemnoidea, and other invertebrates died out; the group of bony fishes also considerably changed. In particular, at the boundary between the last epoch of the Cretaceous (Maastrichian) and the first epoch of the Tertiary (Danii), only 8 of 38 genera and 11 of 50 species were preserved (Development and change... 1978).

The fact of the extinction of many land, water, and flying reptiles at the end of the Cretaceous is widely known; dinosaurs, for example, were among the land reptiles now extinct.

It seems obvious that the most accurate data as to the duration of the late Cretaceous extinction can be found in the most extensive source of paleontological information, the marine vertebrates, in particular, the foraminifera. One study (Smith and Hertogen 1980) on variations in plankton at the Cretaceous-Tertiary period boundary draws the conclusion that the extinction occurred over a period of not more than 200 years. However, other studies based on analogous data claim that the extinction of plankton lasted for a far longer period.

Two reviews of paleontological data treating the question of the duration of the late Cretaceous extinction, recently published in the *Scientific American*, deserve attention. The first concludes that the extinction of various organisms, including dinosaurs and many reptiles was momentary, being caused by asteroid fall (Russell 1982). The second review proposed that this extinction lasted for no less than 2 million years and was induced by climate cooling, the cause of which is not specified (Stanley 1984).

Since these two reviews were written by well-known scientists, the discrepancy in their views reflects the difference in opinions among large groups of paleontologists, and it appears probable that the available paleontological materials do not allow a reliable determination of the duration of mass extinctions. In this connection this question should be considered when discussing the mechanism which caused mass extinctions of organisms.

Causes of Mass Extinctions

Let us dwell first on the problem of the ecological mechanism of mass extinctions, which has been insufficiently taken into account when discussing this question. As noted above, the same climatic change can exert a very different effect on animate nature, depending on the time during which it develops. An example was given of the appearance of Quaternary glaciations, when the temperature was reduced by tens of degrees over vast areas. Since the corresponding climatic change developed in the course of many thousand years, mass extinctions of organisms did not occur. Here it is necessary to point out another reason why mass extinctions were impossible while glaciers advanced. Drastic climatic changes during the ice ages took place only in middle and high latitudes, which allowed many animals and plants to move to warmer regions at lower latitudes, where climate had changed only a little.

It is obvious that if glaciers advanced over the entire Earth's surface (which would theoretically be possible, with appropriate variations in heat influx to the troposphere), all living organisms would become extinct, i.e., the biosphere would be destroyed. This conclusion is confirmed by the fact that living organisms are completely absent in the central Antarctic, which is covered with an ice sheet.

In the previous section an example was given of the considerable decrease in the total number of animal species at the beginning of the Mesozoic, corresponding with the considerable lowering of oxygen in the atmosphere. This decrease was unavoidable, since the change in atmospheric chemical composition was global.

All these facts can be used to judge the hypothesis proposed by Stanley (1984) that the late Cretaceous extinction lasted for 2 million years and is explained by climate cooling.

We can agree with Stanley's conclusion that mass extinction was caused by temperature lowering in the environment. At the same time, it is quite improbable that such cooling developed gradually and lasted for 2 million years.

If this cooling was comparatively weak, i.e., the mean temperature decreased by several degrees (there are some paleotemperature data confirming this possibility), then during millions of years a considerable part of the organisms could have adapted to this cooling by means of migration to warmer regions, acclimatization, or evolutionary modifications. A study of mean temperature variations throughout the Phanerozoic epochs indicates that although in a number of cases these variations comprised several degrees, no mass extinctions occurred with these slow temperature changes.

At the same time it is impossible to assume that during the last million years of the Cretaceous the Earth's mean surface temperature decreased considerably (e.g., by 10°C). Such a cooling would have been discovered in paleotemperature data. In addition, it would inevitably have led to vast glacial advances in the coldest regions, which did not occur at the end of the Cretaceous.

There exists a simple ecological principle of the influence of unfavorable factors on animate nature: the more rapid the action of an unfavorable factor (in this case, the lowering of environmental temperature), the greater the damage caused to organisms.

Therefore, a drastic, even if insignificant, lowering of the Earth's mean temperature over a few years could cause mass extinctions, but these extinctions would not have taken place if the temperature had decreased gradually over many thousands or even millions of years.

With the question of the ecological mechanism of mass extinctions, of considerable importance is the problem of the factors that could have caused these critical epochs of geological history.

Section 1.2 showed the results of studies which established that after explosive volcanic eruptions, due to decreasing atmospheric transparency, the surface air temperature is reduced all over the globe (or in one of the hemispheres). After a single eruption, this reduction in temperature reaches several tenths of a degree; however, after a series of eruptions, the mean air temperature could have been lowered by 5°C or even more. Similar events can be repeated within 10 or 100 million years, which agrees with the time intervals between critical epochs.

The hypothesis that critical epochs are related to collisions of the Earth with comets, large meteorites, and asteroids has been proposed time and again.

As noted above, the hypothesis that the falling of comets on the Earth's surface resulted in the extinction of organisms was proposed by Laplace at the end of the 18th century. At present this hypothesis is supported by a number of authors, including the well-known chemist Urey (1973), who assumed that mass extinctions of organisms can be attributed to the Earth's colliding with comets.

There are indeed grounds to believe that the fall of celestial bodies on the Earth is the factor causing mass extinctions. Observations of the Moon showed that there are many craters on its surface, formed when more or less large celestial bodies fell on it. Cosmic studies of the 1960s–1980s showed that the surfaces of Mars, Mercury, Venus and satellites of Jupiter and Saturn have a similar structure.

It is beyond doubt that no fewer celestial bodies fell on the surface of the Earth than on the surface of these other planets, and that this was the cause of many craters. It is assumed that impact craters (astroblems) 100 km in radius appeared on the average once in 14 million years, those with radius of 500 km once in 600 million years. Although traces of these craters on our planet have in most cases been smoothed out by atmospheric and hydrospheric processes, some astroblems are still preserved (Geology of astroblems 1980).

It is clear that such grandiose phenomena as collisions of the Earth with large celestial bodies must have had an enormous impact on animate nature. They were accompanied with explosions, whose energy was enormous. After the explosion, localized short-term air temperature increase occurred and was then replaced by longer global cooling due to the considerable increase in the atmospheric aerosol layer composed of explosion products. The physical mechanism of such a cooling is probably very similar to that after a single volcanic eruption, although temperature lowering in this case could have been far greater because of the enormous explosive power. It might be thought that after the falling of a comparatively large body, as after an explosive volcanic eruption, the atmosphere accumulated a great amount of aerosol particles of explosive origin.

Attempts have recently been initiated to find scattered meteorite matter in the layer that synchronizes with epochs of great extinctions of animals, and have given interesting results. In the layers belonging to the end of the Cretaceous, an excessive amount of iridium has been discovered, which is typical of extraterrestrial bodies. Later, it was established that other metals belonging to the platinum group, as well as nickel and cobalt, showed similar concentrations. This phenomenon was discovered not only on various continents and islands, including Europe, Africa, North America, and New Zealand, but also at the bottom in the central Pacific, proving its global nature.

From data on the amount of iridium in sediments, the sizes of the extraterrestrial body that fell on the Earth have been estimated to be 5–16 km, with a mass of the order of 10^{14} kg. Since these values are typical of asteroids, the celestial body that collided with the Earth was thought to have been an asteroid. The kinetic energy of this collision comprises about 10^{23} Joules, approximately equivalent to the explosion of 10^{14} tons of trinitrotoluene (Alvarez et al. 1980; Ganapathy 1980; Hsü 1980; Newell 1980; Smit and Hertogen 1980; Hsü et al. 1982; O'Keefe and Ahrens 1982). In studying this problem, different hypotheses have been proposed as to the after effects of the fall of the asteroid which caused animal extinctions at the end of the Cretaceous. For example, it has been assumed that after this fall the amount of dust thrown up into the stratosphere exceeded by three orders of magnitude the amount of aerosol raised to the stratosphere after the largest explosive eruption of the last 100 years (Krakatoa, 1883). It has been assumed that for several years the stratospheric dust screen could have allowed no solar radiation on to the Earth's surface. This resulted in a cessation of photosynthesis and mass extinction of animals (Alvarez et al. 1980). Other authors consider that the extinction was associated with the heating of the atmosphere that occurred either just after the fall of the asteroid or in the thousands of years following, due to the atmospheric accumulation of carbon dioxide, which could be less absorbed in the ocean because of the death of the plankton induced by cessation of photosynthesis (Hsü et al. 1982).

These assumptions about the mechanism of the mass extinction at the end of the Cretaceous do not appear to be true. Cessation of photosynthesis on the entire planet for several years would have caused the extinction of almost all life, in particular the organisms in the oceans, where only very poor nutritive resources of organic matter exist. Since after the mass extinction many oceanic organisms were still alive, the assumption would appear to be based on a considerable overestimation of the amount of dust ejected to the stratosphere after the fall of the asteroid. One can also believe that this fall was not accompanied with considerable heating of the entireatmosphere. If the asteroid fell into the ocean, the conversion of the energy of its fall into heat would not have led to a significant increase in ocean temperature. Besides this, a major portion of the energy of the fall would have probably been absorbed by the Earth's crust, in which the asteroid was buried deep.

The heating of the atmosphere as cause of animal extinction also seems highly improbable, because of the selective character of the extinction. Vertebrates, including many groups that had no thermoregulation, disappeared, whereas warm-blooded animals (mammals and birds) did not suffer. This indicates the extinction of animals at the end of the Cretaceous to have been induced by a short-term cooling.

Calculations show (see Sect 1.2) that at that time the temperature drop was 5–10°C. This would undoubtedly be sufficient for the extinction of more or less stenothermal organisms, which were numerous at the end of the Mesozoic, since a warm or hot climate prevailed at all latitudes at that time.

In addition to volcanic and meteorite hypotheses, many theories exist as to the causes for the drastic changes in natural conditions in the geological past, although these are frequently insufficiently corroborated and sometimes contradictory (see reviews by Newell 1967; Herman 1981 etc.). In particular, the possibility of the effects on the earth's organisms of a supernova explosion occurring not far off the Solar System has in several studies been checked against the isotope composition data of a number of elements at the time of mass extinctions, but with negative results (The quest... 1980).

The question of possible mechanisms producing drastic changes in the natural environment is not yet solved and requires further study. Keeping in mind the narrowness of the "climatic zone of life" (the range of atmospheric conditions within which organisms can survive), it is highly probable that for the long time during which the biosphere has existed, for individual groups of organisms this zone has disappeared at certain moments as a result of various environmental factors, including those above. The most important example of such a change in natural conditions is the ecological crisis at the end of the Mesozoic era.

We can come to the conclusion that the cooling that induced this crisis was not very strong. The inference drawn in some calculations, that when the asteroid fell, the mean air temperature over the continents was lowered by 40°C and for half a year remained below freezing point, is open to question. Such a cooling would have led to the extinction of a far greater number of continental flora and fauna species than actually disappeared at the end of the Cretaceous. The possibility of decreasing solar radiation for several years below the limit at which photosynthesis can occur, as Alvarez et al. (1980) believed at first, also seems to be exaggerated. In their later paper, this period of time was reduced to several months (Alvarez et al. 1982), Toon et al. (1982) have come to a similar conclusion. We think that this estimate is probably also exaggerated, and that in addition to a decrease in photosynthesis intensity, the lowering of the water temperature by several degrees was a significant cause of the extinction of marine organisms. As mentioned above, animals at the end of the Mesozoic could have been very sensitive to even insignificant temperature variations.

The mass extinction at the end of the Cretaceous is one of several large-scale extinctions throughout the Phanerozoic. It is possible that the extinction at the end of the Permian was on an even larger scale.

During the Phanerozoic several large extinctions were undoubtedly accompanied with a far greater number of smaller events of the same nature, which it is difficult to discover because of the incompleteness of paleontological history. Available data show that some of these less significant extinctions were also caused by the falling of external bodies onto the Earth (Asaro et al. 1982; Ganapathy 1982; McGhee 1982; Palmer 1982).

In the last few years, the problem of mass extinctions has been discussed at a number of scientific meetings. One of them, held in the U.S. in October 1981, dealt with the effects of the fall of large external bodies on the extinction of organisms. The report of the meeting (Simon 1981) emphasizes that in discussing this problem the possibility was recognized of a causal relationship between the indicated events. It is interesting that this recognition was considered to be a "scientific revolution."

A broader problem was discussed at a meeting in West Berlin held in May, 1983, devoted to clarifying the role of "abrupt events" in the Earth's history. As can be seen in the review (Fifield 1983), several reports at this meeting noted that it is impossible to understand many phenomena of the Earth's history based on the concept of uniformitarianism. At the meeting, the role of large volcanic eruptions as a factor causing the extinction of organisms was discussed and the importance for mass extinctions of the falling of external bodies was emphasized. The reports presented contained the results of calculating the probable frequency of the collision of asteroids and comets of various sizes with the Earth, and new estimates of the frequency of mass extinctions in the Phanerozoic. The good agreement between the estimates of these frequencies deserves our attention.

Recently, many studies have appeared confirming the fact of mass extinctions in the Earth's history and establishing the relationship between these extinctions and drastic changes in the abiotic environment. Although the question of the actual mechanism of mass extinctions is far from being completely solved, the greatest probability is that at least a part of these extinctions is associated with comparatively short-term changes in natural conditions after the fall of large external bodies on the Earth's surface and after large volcanic eruptions. One more example confirms this assumption: McLaren (1970, 1982) presented information showing the mass extinction of marine fauna in the middle of the Late Devonian (365 ma ago). Among possible causes of this extinction, McLaren indicated the fall of a large meteorite. Recently, in Australia, elevated concentrations of iridium have been discovered in the layer corresponding to the Devonian extinction (Playford et al. 1984; McLaren 1985a,b).

In conclusion, several additional ideas can be presented concerning the question of the duration of mass extinctions.

Although the hypotheses presented here concerning the causes of the extinctions appear conclusive as far as the momentary impact of aerosol climatic catastrophes on animate nature is concerned, they give no unambiguous answer to the question of the duration of these extinctions. Even in the first years of studies of aerosol catastrophes, it was noted that the extinctions could be induced not by one unusually large volcanic eruption (or a series of eruptions following closely in time) but by a series of large eruptions that occurred in the epochs of maximum volcanic activity, separated by time intervals of many thousands of years. First coolings could have killed animals less adapted to a cold climate, and the subsequent ones a number of other groups weakened by previous coolings (Budyko 1971).

It was later proposed that the falling of an asteroid might have initiated the dislocation of magma masses in the lithosphere and caused a more or less prolonged increase in volcanic activity. This could have resulted in the appearance of particularly large volcanic eruptions. Thus, after an initial aerosol catastrophe, a number of new catastrophes could have taken place that caused the extinction of the organisms that had survived the fall of the asteroid (Budyko 1982).

An interesting but disputable hypothesis has recently been proposed that the fall of external bodies on the Earth's surface could have led to a series of aerosol catastrophes, separated by prolonged time intervals.

Raup and Sepkosky (1984) provide new data on mass extinctions in the Phanerozoic. These extinctions were smaller than the largest extinctions discovered in their previous study. They came to the conclusion that these extinctions occurred at approximately equal time intervals of about 26 million years, a conclusion that caused Miller and other American astrophysicists to take an interest in the problem. They have assumed that these extinctions could have been induced by a small star, a satellite of the Sun, which has not yet been discovered by observation. They believe that this star, called Nemesis, revolves in its elliptic orbit at an average distance from the Sun which is 100,000 times greater than the distance between the Sun and the Earth. This would mean that the time of rotation of Nemesis around the Sun equals 26 million years, during which, from time to time, Nemesis approaches the Oort cloud, the region containing comets and remnants of spent matter in the formation of the Sun and the planets of the Solar System. The influence of Nemesis on the Oort cloud then causes mass ejections of comets towards the Sun, some of which then collide with the Earth. The duration of one stage of formation of such comets is considered to be close to 10^6 years (Thomsen 1984).

This hypothesis gives rise to doubt for two reasons: first, the failure to prove the statistical reliability of the periodicity of mass extinctions established by Raup and Sepkosky. The conclusion about the periodicity is drawn from data on a comparatively small number of cases of extinctions, and the possibility of proving each of these cases and the exactness of the estimated age (on which the fact of periodicity depends) is limited.

Second, the assumption of the long existence of a small star, the Sun's satellite, raises objections from many astronomers, who believe that this small star could not have been a part of the Solar system for a long time, since gravitation is weak at such a great distance from the Sun. The discussion of this question has continued up to the present (Weisburd 1984).

Taking into account that the question of the duration of mass extinctions has not yet been completely solved, we note that under the circumstances there is a great probability that aerosol climatic catastrophes played the determining role in at least some mass extinctions. The conclusion about the effects on the biosphere of aerosol climatic catastrophes in the geological past is considered by many specialists to be one of the greatest discoveries in the field of Earth sciences. The President of the International Union of geological sciences, Dr. E. Zeibold, reviewing the development of geology over the past years, considered the discovery of iridium layers in marine and continental sediments, these indicate the fall of a gigantic asteroid, to be one of the most important achievements of this science. This has allowed us, in his opinion, to clarify the paths of organic evolution on our planet and confirm the fact of mass extinctions of marine and continental organisms (Zeibold 1985). A similar opinion was expressed by the Director of the Geological Survey of Canada, who, quoting a new concept of mass extinctions, pointed out that absolutely new views have appeared concerning biological evolution "...It is interesting that specialists studying this phenomenon are engaged at the same time in considering the aftereffects of nuclear explosion. It was assumed that nuclear collision could result in drastic temperature reduction, first of all because of raising dust into the atmosphere. This points again to the fact that there would be no winners in a nuclear war" (Price 1985). This conclusion is of great importance for the question discussed in the next chapter.

2
Climatic Effects of a Nuclear Conflict

2.1 Climatic Aerosol Catastrophe

MAN-MADE CLIMATIC CHANGE

The discovery of the present anthropogenic change in climate appeared to be one of the most important results of the scientific revolution that has taken place in the field of climatology in recent decades. Figure 11 (curve 1) gives the first published forecast of the mean air temperature changes at the end of the 20th and in the 21st century produced by carbon dioxide accumulated in the atmosphere as a result of carbon fuel combustion (Budyko 1972). In following years, forecasts of this kind were made by a number of scientists and at different scientific conferences, the results of all of them being much the same. To illustrate this, Figure 11 presents curves 2 and 3, which are based on the forecast proposed by the Soviet-American meeting of experts on the climatic effects of increasing carbon dioxide in the atmosphere (Climatic Effects of Increased Atmospheric Carbon Dioxide 1982). Comparison of curve 1 with curves 2 and 3 has shown that the conclusion as to a future large-scale warming, which might occur within the next decades, has in essence not changed, even though great progress has been achieved in the study of anthropogenic climatic change.

Modern studies of climatic change devote great efforts to the problem of detecting this warming, which has already occurred by additional man-made carbon dioxide being introduced into the atmosphere. The analysis of observational data on increasing carbon dioxide concentration has shown that, since the end of the last century, carbon dioxide concentration in the atmosphere of the 1980s has increased by approximately 20%. Climate model calculations show that such a change in the chemical composition of the atmosphere should theoretically have raised the mean air temperature by about 0.5°C. To find how much the global mean air temperature has changed over the past 100 years is difficult, because its variations are disguised to a considerable extent by the effects of fluctuations in the atmospheric transparency discussed in Section 1.1. Although this issue has sometimes been disputed, there are grounds to believe that anthropogenic warming was already traced in the second half of the 1970s, first by Soviet, and then by other scientists (see Budyko and Vinnikov 1983). In some reports of scientific meetings concerned with climatic change, the conclusion about anthropogenic

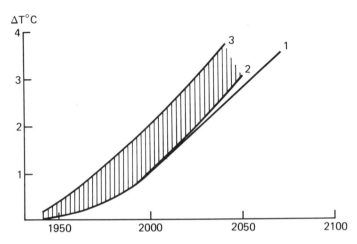

FIGURE 11. Forecasts of anthropogenic climate change.

warming is formulated in the following way: the detected mean air temperature increase of about 0.5°C, which has taken place during the past 100 years, does not contradict the assumption that it was produced by an increase in atmospheric carbon dioxide concentration (e.g., see *International Assessment of the Impact of an Increased Atmospheric Concentration of Carbon Dioxide on the Environment.* WMO/ICSU/UNEP Conference, Villach, Austria, October 1985).

At the same time, it has been found in a number of studies (Alexandrov et al. 1982, Karol 1983) that an increase in concentrations of such minor atmospheric gas constituents as nitrous oxide, methane, tropospheric ozone, hydrofluorocarbons, etc. also leads to climatic warming. These gases absorb little solar radiation, but appreciably enhance the atmospheric opacity for the thermal infrared spectrum, which means that they intensify the greenhouse effect.

The detection of anthropogenic climatic changes produced by increased concentrations of carbon dioxide and certain minor atmospheric constituents is of great importance in understanding the nature of the anthropogenic climatic catastrophe that might occur at present. The fact emerges that man's impact on climate is not creating new, hitherto unknown ways of changing the climate. The physical mechanism of the present warming is principally identical to the mechanism that evoked numerous warmings (as well as coolings) in the geological past, which, as has been shown in Chapter 1, were mainly produced by an increase or decrease in atmospheric carbon dioxide.

The principles governing the initiation of a natural aerosol catastrophe and a possible anthropogenic climatic catastrophe are even closer to each other. The possibility of an anthropogenic climatic catastrophe follows from the studies carried out in the USSR about 10 years ago. The works devoted to this problem use the conclusion given in Chapter 1, that an increase in the mass of optically active particles of stratospheric aerosol by about 10^{-6} g cm^{-2} decreases the temperature of the atmospheric lower layer by 1.5°C if the climatic system is stationary. This means that with a stratospheric aerosol mass in the Northern Hemisphere of about 0.5×10^6 tons, the air temperature will drop by several tenths of a degree. We have reason to think that modern technological means may produce and sustain a layer of optically active atmospheric aerosol mass of up to a few million tons, whose influence on the climate will result in a significant decrease in air temperature (Budyko 1974). This would prevent the development of warming produced by the accumulation of carbon dioxide and other minor gas constituents in the atmosphere. We are not concerned here with the question of whether such measures are expedient, which is not at all obvious.

It can easily be seen that the building of a dense aerosol layer in the absence of (or with very weak) anthropogenic warming will lead to drastic consequences. A similar catastrophe might also occur under a nonstationary state of the climatic system, when a dense aerosol layer generated by a single impact on the physical processes in the stratosphere persists for a limited period of time.

As has been mentioned in Chapter 1, in this case the effects of the changed mass of the aerosol (within several months) on the surface air temperature will be weaker by an order of magnitude than similar effects in the case of a stationary climatic system (Budyko 1971). This means that dramatic consequences will arise when in one of the hemispheres the mass of optically active aerosol particles increases by several tens of millions of tons (the estimates of the climatic effects of the volcanic eruptions given in Chap. 1 are almost the same).

The problem of the possible climatic effects of a nuclear war was raised in 1975 in a report prepared by the US National Academy of Sciences (US National Academy of Sciences 1975). It was stated that the explosion of the warheads available in the weapons at that time would result in the injection of 10^8 tons of dust into the upper atmosphere, which is comparable with the mass of aerosol released into the atmosphere by the Krakatoa volcano. The authors of the report came to the conclusion that the aftermath of such an explosion cannot significantly change global climate and biosphere.

The studies published before 1975 clearly demonstrate that this conclusion is not valid. Assuming that the estimate of the total aerosol mass released into the atmosphere given in the report is reliable, we should remember that not all of this aerosol mass can be a climate-forming factor because it does not only consist of optically active particles that have reached the stratosphere. The volume of these particles will at the same time be supplemented by the combustion products from the fires which will spread over vast areas following nuclear explosions. Until 1982, this factor was not taken into account. Considering that in the first approximation these two factors will compensate for each other, we find that on the average the mass of optically active aerosol in the Northern Hemisphere will be about 0.4×10^{-4} g cm^{-2}. This mass of aerosol particles could decrease the mean surface air temperature by about 10°C. The temperature decrease over the continents will be much greater than the mean global one. Such a cooling will undoubtedly bring drastic consequences.

This very simple estimate of climatic change was reported by one of the present authors in 1982 and published somewhat later (Budyko 1984). Since this conclusion was based on the early studies of the late 1960s-early 1970s and does not involve detailed model calculations, we have presented it before proceeding to the results of thorough investigations of the climatic effects of a nuclear war.

It was assumed in this first study that the mass of optically active aerosol particles injected into the stratosphere after a nuclear conflict might be hundreds of millions of tons, therefore the initial aerosol loading in a vertical column of the stratosphere will be of the order of 10^{-4} g cm^{-2}. It follows from the calculations by the formulas of atmospheric optics that such a concentration of stratospheric aerosol will reduce by more than half the amount of solar radiation reaching the troposphere.

Taking into account the air temperature/solar radiation dependence, we find that, with a stationary state of the climatic system, the mean surface air temperature might drop by several tens of degrees. The absolute values of the actual decrease will be appreciably smaller because of the thermal inertia of the climate system. This influence can easily be calculated by using simple empirical formulas. It can then be seen that with a residence time of the stratospheric aerosol of several months, the mean air temperature decrease near the Earth's surface will be 5° to 10°C. This estimate coincides with the value, given earlier, of temperature decrease after the impact of a major asteroid on climate.

It should be noted that the potential temperature decrease in different regions of the Earth can greatly deviate from this average value. Since in the Northern Hemisphere the aerosol mass will increase much more and the ocean area is smaller than in the Southern Hemisphere, the expected mean air temperature decrease in the Northern Hemisphere will be approximately 15°C and over the continents of the Northern Hemisphere it could be more than 20°C.

A clear idea of the spatial temperature distribution after a nuclear war can be drawn from a schematic map of the possible distribution of decrease in air temperature several months after a nuclear war (Budyko 1985). The value of decrease in solar radiation taken to construct the map (Fig. 12) was smaller than that given in the above calculation. The data for the map were obtained from empirical information on annual temperature variations, so that in some respects they are more reliable than the results obtained by model calculations. Temperature anomalies given in Figure 12 correspond to a certain sequence in the reduction of radiation influx to the troposphere (i.e., its seasonal variations), which can differ greatly from the changes in radiative fluxes after a nuclear war. The latter circumstance is possibly not of great importance, because these scenarios are very conditional, and therefore any calculations of the spatial distribution of possible climatic changes after a war are of preliminary character. It is more important for estimating the reliability of this scheme (as well as of other schemes of the spatial distribution of temperatures following nuclear conflict) that modern studies cannot take into account the effects of highly nonstationary atmospheric processes on climatic change, which is discussed below in more detail.

The most abrupt cooling will probably continue for a few months, and a lesser cooling will last appreciably longer. Without dwelling on other aspects of possible climatic changes after a nuclear war, let us note that these changes will resemble

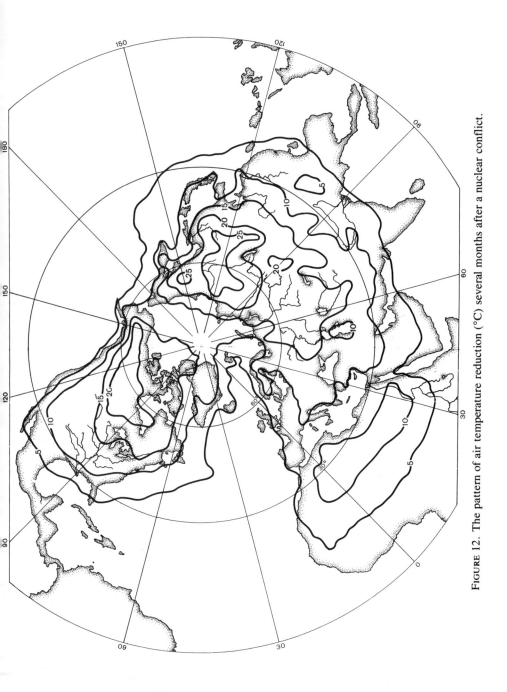

FIGURE 12. The pattern of air temperature reduction (°C) several months after a nuclear conflict.

a major climatic catastrophe comparable to those of the geological past which were discussed in Chapter 1.

These calculations of the possible climatic change after a nuclear war have been very rough, which, in our opinion, can be justified by the great uncertainty in a number of the parameters used, particularly of those for estimating the amount of aerosol injected into the stratosphere. In this connection we may state here our opinion that the estimates obtained suggest the possibility of catastrophic effects on the entire biosphere in the event of major nuclear warfare.

This catastrophe will not only sweep away numerous species of animals and plants, but bring new distress to the fraction of mankind that will survive the immediate nuclear conflict. It is very likely that the abrupt cooling (particularly in areas of warm climate, where the houses are not heated) will prove fatal for many people, who will die of cold. However, this will not be the capital danger with which mankind will be faced by the worldwide cooling. More severe losses will come from the failure of agricultural production. Such a drastic drop in mean temperature will lead to an appreciable reduction in photosynthetic productivity, which may result in a complete or almost complete elimination of crop yields. As has been indicated by Crutzen and Birks (1982), photosynthesis may decrease also because of a substantial reduction in the sunlight reaching the Earth's surface.

Since the available global food reserves are very small compared with the annual food production, elimination or a sharp reduction of food production will take the lives of many people (this problem is thoroughly treated by Harwell and Hutchinson 1986).

Because the accuracy of the calculation presented here is limited, the question of aerosol climatic catastrophes in the geological past appears to be important to confirm the reality of a possible aerosol climatic catastrophe following a nuclear war. The possibility of inducing such phenomena by natural factors and their strong impact on the biosphere considerably increases the probability that a similar catastrophe can happen after a multiple nuclear exchange.

Studies Concerned with the Climatic Effects of Nuclear Conflict

Let us describe in more detail the development of these studies. In 1975, the US National Academy of Sciences issued a report which came to the unsubstantiated conclusion that a serious climatic change cannot be expected after a nuclear war.

The work of Crutzen and Birks (1982) proves to be important in studying this problem. It was the first attempt at a quantitative evaluation of the input to the atmosphere of sooty smoke generated by the massive fires which might be ignited by nuclear explosions. Their estimates have shown that the smoke cover will drastically attenuate the solar radiation flux. The authors of the paper have noted the danger of a considerable reduction of photosynthesis, and mentioned the possibility of climatic change, however, without any quantitative estimates.

In its later report (1982), the US National Academy of Sciences indicates that work on the problem of climatic effects of a nuclear war were started in the USA in 1982. The first publications of American scientists appeared at the end of 1983.

In the USSR, the possibility of aerosol catastrophes was suggested in 1969, and the idea of possible realization of man-made aerosol climatic catastrophe followed from the work by Budyko published in 1974.

In 1982-1983, in reports at different conferences, the present authors drew attention to the possibility of the disastrous worldwide effects of a nuclear war.

On May 9, 1983, Izrael made a report devoted to this problem at the 9th Congress of the World Meteorological Organization, in discussing the issue *Meteorology and Society* (Izrael 1983a).

Obukhov and Golitsyn presented a paper *Possible Atmospheric Effects of a Nuclear War* at the Conference of the Soviet Scientists, *For Peace and Deliverance of Mankind from Nuclear Threat*, held in Moscow on May 16-18, 1983. It was stated in this paper that the temperature of the lower atmosphere and the underlying surface, which are highly contaminated by smoke or dust, has to approach thermal equilibrium, determined by a balance between the solar energy absorbed by the atmosphere and outgoing thermal radiation of the Earth's surface and the atmosphere. Under normal conditions, this temperature is equal to 255°K = −18°C. It was also noted there that a strong attenuation of the hydrologic cycle under these conditions should make for longer residence time of aerosols in the atmosphere, and it was indicated that the general atmospheric circulation would inevitably change. All this has been stated in more detail in a study carried out within the framework of the *Committee of the Soviet Scientists for Peace, Against Nuclear Threat* (Golitsyn and Ginzburg 1983). In 1983-1984 there appeared the results of the first studies of the Computing Center of the USSR Academy of Sciences concerned with numerical modeling of climatic consequences after the injection into the atmosphere of large quantities of smoke (Alexandrov and Stenchikov 1983, 1984). In 1983, the works of Izrael raised the questions of the destruction of the ozone layer, of the geophysical and ecological effects of a nuclear war. His monograph reviewed this problem as it was known by the beginning of 1984.

In autumn 1983, the Conference on Biological Effects of a Nuclear War was held in Washington. It was at this conference that Sagan and Turco introduced the term "Nuclear Winter". The results of calculations by general circulation models carried out at the Computing Center of the USSR Academy of Sciences and at the US National Center for Atmospheric Research presented at this conference proved to be very close to each other. It appeared that, following a nuclear war, the inland temperature of the underlying surface will drop appreciably (the temperature will be much below freezing point). Similar results have been obtained by a simple, spatially averaged, so-called radiative-convective model by Turco et al. (1983) and by a still simpler model of Golitsyn and Ginzburg (1983, 1985).

Immediately after the Washington conference, a scientific discussion was held on television (Washington-Moscow TV-bridge), in the course of which Soviet and American scientists discussed various potential effects of a nuclear war, such as biological, atmospheric, ecological, and medical aspects. Izrael, one of the authors of this book, took part for the Soviet team in discussing the atmospheric and ecological impact of a nuclear conflict. He also presented a paper on the same subject at the 9th International Conference on Cloud Physics that was held in Tallin in August, 1984.

Since the end of 1983, the Scientific Committee of the International Council of Scientific Unions on the Problem of Environment, SCOPE, has started active work on nuclear impact on nature, biosphere, and agriculture. On the basis of the ENUWAR (Environment and Nuclear War) Project a number of scientific seminars and workshops were held, with scientists from different countries (one of the seminars was held in Leningrad in May, 1984 and another in Tallin in August, 1984), as a result of which a report in two volumes was prepared and published: Volume 1: *Physical and Atmospheric Effects* (Pittock et al. 1986) and Volume 2: *Biological Effects of Nuclear War*, including agricultural effects (Harwell and Hutchinson 1986).

On the initiative of the USSR, in 1984 WMO asked the Joint Scientific Committee of WMO/ICSU, which directs the World Climate Research Program, to discuss the possible climatic consequences of a nuclear war. The members of this Committee, G.S. Golitsyn and N.A. Phillips, prepared a report *Possible Climatic Consequences of a Major Nuclear War*, which was approved and presented to the WMO Executive Council.[1] Some aspects of this report will be discussed in Section 2.3.

At the present time in the USSR, the USA, Australia, West Germany, Great Britain, Canada, and in some other countries, extensive work is being done to reassess the effects of a nuclear war on the atmosphere, climate, and the biosphere as a whole. Despite uncertainties in solving a number of problems, there is no doubt at all that the catastrophic worldwide impact of nuclear conflict is absolutely feasible.

The results of all the studies in this field show that nuclear war should be precluded in every possible way, because it will inevitably lead to a large-scale disturbance of the climate on our planet and inflict irretrievable damage to the entire biosphere and to mankind as a whole.

Let us now discuss some of the published scenarios for a nuclear war. All scenarios available at present, i.e., the Ambio scenario (1982), about 20 scenarios of Turco et al. (1983), a scenario from the indicated report (US National Research Council 1985) and the ENUWAR scenario (Pittock et al. 1986) have been devised either by individual scientists or by non-governmental organizations, and therefore they reflect no official view. At the same time, they have some common features, which we shall discuss. For example, all scenarios assume that, in attack and response, both sides can direct their strike not only

---

[1] In 1987, an update of this report was presented to a Joint Scientific Committee and WMO (see Golitsyn G.S. and M.C. MacCracken. Atmospheric and climatic consequences of a major nuclear war: results of recent researches. WCP-142, WMO, Geneva, December 1987).

against key military targets but also against the industrial and economic potential of an adversary, most of which lies in and near the cities.

According to these scenarios, the underground targets will be attacked by massive surface bursts with an explosion yield of 1 Mt to about 20 Mt and the surface targets will be destroyed by 50 to 500 kt explosions.

In comparison, we recall that in August, 1945, Hiroshima experienced the explosive yield of an atomic bomb of about 12 kt of the TNT energy equivalent, and Nagasaki 20 kt. Altogether there are now in the world up to 50,000 nuclear warheads (Pittock et al. 1986) with the total yield of about 12,000 Mt. The basic scenarios assume the usage of about half the amount. It was also assumed that about half the total number of explosions will be surface bursts.

The surface bursts and near-surface explosions (at a height below 0.5–1.0 km) will produce deep craters and an immense quantity of debris (dust).

With air bursts where the fireball is not in close contact with the ground (Glasstone and Dolan 1977), the principal damaging factors are the thermal pulse (light emission) and blast wave.

The thermal pulse of a nuclear airburst is intense enough to ignite extensive urban and particularly forest fires, as well as oil and gas fires. Massive fires could generate large quantities of smoke and various gaseous substances that would rise into the atmosphere. Aerosol particles produced by the explosion and fires reduce the atmospheric transparency and change other radiative properties of the atmosphere.

As can be seen from Table 2 (Izrael 1983b), most of large-scale effects of nuclear explosions lead in the long run not only to changes in weather and climate, but to other serious geophysical consequences.

The surface nuclear bursts produce enormous quantities of dust; up to 5000 tons of rock debris are ejected into the atmosphere by an explosive yield of 1 kt, a small portion of which, up to 15–25 tons/kt, becomes vaporized (Izrael 1973, 1984). The vaporized portion is then converted into finely dispersed aerosols with an average particle size of fractions of a micrometer (surface nuclear bursts of 2500 Mt yield about 40 to 60 Mt of such material).

Most of the ejected rock and soil fall back to the ground near the explosion site; a major portion of vaporized matter (up to 80%) is caught by coarse-dispersion aerosols and the remaining finely dispersed aerosol particles fall out to the Earth's surface within weeks, months, or even years. The total mass of ejecta entrained into a radioactive cloud decreases rapidly with increasing height of detonation above the surface. Air bursts at high altitudes produce an aerosol mass which is almost equal to the mass of a nuclear device.

In the lower atmosphere the fireball rises at a rate of about 100 meters per second. The rising fireball grows in size, turns into a torus and forms a cloud, gradually losing its buoyancy as it reaches the more rarefied atmosphere. Finally, its mass approaches the mass of displaced air and it then stabilizes at a certain height. The height of elevation depends on the burst power and the atmosphere's stratification. Figure 13 shows (Peterson 1970) the positions of the upper and lower boundaries of the cloud, depending on the explosion yield for the tropical

TABLE 2. Geophysical (ecological) effects of principal large-scale damaging factors of nuclear explosions

Principal large-scale effects (damaging factors)	Possible geophysical consequences
1. Biosphere contamination with radioactive fallout	Changes in electrical properties of the atmosphere, weather change Changes in the properties of the ionosphere
2. Pollution of the atmosphere with aerosol particles	Changes in radiative properties of the atmosphere Weather and climate changes
3. Pollution of the atmosphere with various gaseous products (methane, ethylene, tropospheric ozone, etc.)	
Pollution of the troposphere	Changes in radiative properties of the atmosphere, changes in weather and climate
Pollution of the upper atmosphere	Changes in radiative properties of the upper atmosphere, perturbation of the ozone layer Changes in the flux of ultraviolet solar radiation, climate change
4. Changes in the Earth's surface albedo	Climatic change

atmosphere ($<30°$ lat) and the atmosphere of higher latitudes ($>30°$ lat). In low latitudes where the tropopause is found at higher altitudes, the fireball rises higher. However, in relation to the position of the tropopause itself, the difference is small between the heights reached by a fireball in low and high latitudes. In the case of mass nuclear air bursts, the total amount of the inert material converted into finely dispersed aerosol particles will evidently be limited to tens of thousands of tons. Given a significant contribution from the surface explosions, this amount might increase by one or two orders of magnitude. The ejected material will be raised to the stratosphere to a height of 10 to 40 km, and will remain there for a long period (for months or even years).

To estimate the altitudes of the dust rise in the atmosphere and the dust characteristics, one can use data on the nuclear aerosol burden from Izrael's book *The Isotopic Content of Radioactive Fallout* (1973) and Glasstone and Dolan's *The Effects of Nuclear Weapons* (1977).

In a surface nuclear detonation, the size distribution of the dust particles entrained into the nuclear cloud can be described by log-normal law (Izrael 1973):

$$N(d_1,d_2) = \frac{1}{\sigma\sqrt{2\pi}} \int_{\lg d_1}^{\lg d_2} e^{-\frac{(\lg d - \lg \bar{d})^2}{2\sigma^2}} d(\lg d), \quad (1)$$

where d is the particle radius in μm and σ is the dispersion. For surface bursts in siliceous soils, $\lg \bar{d} = 2.053$ and $\sigma = 0.732$, while for bursts at the coral ground $\lg \bar{d} = 2.209$ and $\sigma = 0.424$.

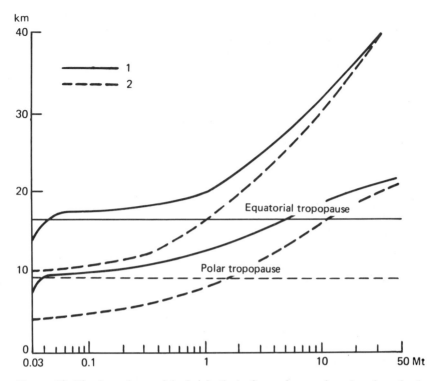

FIGURE 13. The dependence of the height (km) of a mushroom-shaped nuclear cloud on the explosive yield (Mt). 1 = The top and bottom of the cloud in the tropics; 2 = in middle and high latitudes.

The character of this distribution is very close to the distributions of the mass, volume, and activity of the particles, and consequently, it can also be applied generally to describe the nuclear material entrained into the cloud. Large particles are distributed according to the power law with the exponent close to -4.

According to (1), the mass of the particles less than 1 μm is equal to 0.2% of the total particle mass in the cloud. Then the mass of the particles of this type is 50 Mt, assuming aerosol production of 5000 tons per 1 kt of explosive energy and total bursts of 2500 Mt.

According to Turco et al. (1983), the total dust amount in the cloud (based on the total yield of 5000 Mt) is 9.6×10^8 tons; 80% of this value reaches the stratosphere, the portion of particles smaller than 1 μm being 8.4%. Thus, the total mass of finely dispersed aerosols in the stratosphere will be 80 Mt.

For comparison, it can be mentioned that a month after the El-Chichon eruption, the stratospheric aerosol burden (with particle size less than 1 μm) was 23 Mt, and in six months 8.3 Mt (Kondratjev 1985). Let us recall that the Krakatoa eruption produced about 30 Mt of aerosols of all sizes, but with a somewhat different particle size distribution.

2. Climatic Effects of a Nuclear Conflict

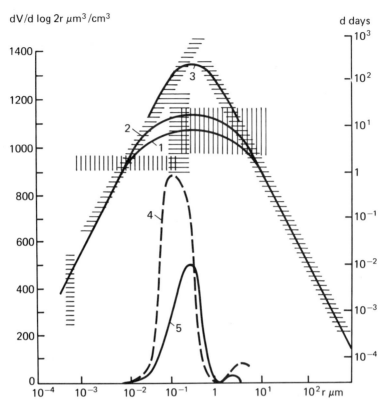

FIGURE 14. The dependence of aerosol residence time (d) on the particle radius (r) at different altitudes (Jaenicke 1981). *1* = Below 1.5 km; *2* = in the mid-troposphere (about 5 km); *3* = near the tropopause (10–11 km); *4* and *5* = distribution of smoke particles by volume depending on their radius (Stith et al. 1981) for a sample taken at a height of 1.83 km and a distance of 10 km downwind from the smoke source *4* and at a height of 1.22 km and a distance of 18.5 km downwind *5*. A range of possible variations in the parameters is represented by the *hatching*.

According to the principal scenario of the U.S. National Research Council Report (1985), about 10 to 23 Mt of the dust particles in the size range below 1 μ could be lifted into the stratosphere. The difference arises because this scenario, apart from that of Turco et al., assumes a greater number of less powerful explosions near the ground. At the same time, about 200 to 500 Mt of nuclear dust particles might penetrate into the troposphere, the submicron fraction making up 8% (i.e., 17 to 40 Mt). The dust can fall out of the lower troposphere in about a week; however, the residence time of dust particles in the upper troposphere will be several times greater.

The same report considers a case when a hundred 20 Mt hydrogen bombs will be expended. According to Figure 14, the nuclear cloud produced by such an explosive yield will stabilize at a height of 19 to 37 km, which means that practically all the dust raised from the Earth's surface will come into the stratosphere. With the most probable uptake of 0.3 Mt of dust per 1 Mt of explosive energy, the stratosphere receives 600 Mt of dust (the dust quantity might vary from 200 to 1000 Mt), about 50 Mt being a submicron fraction. The lifetime of dust particles in both the stratosphere and the troposphere depends on the particle size.

Before considering the problem of how the nuclear dust raised into the atmosphere affects the transmission of solar radiation, we shall present here more detailed information on the residence time of aerosol particles in the atmosphere and on their refraction index. The residence time of the particles depends also on the altitude at which they stay. The average residence times of aerosol particles in the troposphere are given in Figure 14 (Jaenicke 1981).

All the curves exhibit a more or less sharp peak in the particle size range between 0.1 and 1 μm. Figure 14 shows that at a height of 1.5 km, the particles in this size range can survive for approximately one week, while in the lower stratosphere near the tropopause they can stay for many months (up to a year), although these values can differ greatly under various actual conditions. The lifetime of particles smaller than 0.1 μm is shorter, because they tend to coagulate with larger particles. The particles greater than 1 μm are more active centers of condensation, and are washed out by precipitation (wet deposition). The larger particles, more than 10 μm in size, can fall out of the atmosphere rather quickly due to the force of gravity (gravitational deposition). In addition to this, in a turbulent motion of air masses containing practically nonsettling material, part of this will settle on the Earth's surface as a result of sorption, i.e., by so-called dry deposition, at an effective rate of 0.5 to 0.8 cm/s.

Other things being equal, the residence time of aerosol particles in the stratosphere is much longer than that in the troposphere.

The refraction index is the main optical characteristic of the aerosol. Pollack et al. (1973) have measured a complex refraction index for volcanic and other rocks. For the visible waverange, the real part of the refraction index for these rocks appeared to be between 1.47 and 1.57, while the imaginary one varied from 2×10^{-5} (volcanic glass) to 1×10^{-3} for other rocks (andesite). As a result, the refraction index for dust in the visible range of wavelengths, $m = 1.5 - 0.001i$ is most often used. In the thermal range, the optical properties of these materials appreciably depend on the wavelength, since around 10 μm silicon, which the rocks mainly consist of, has strong absorption bands. In the atmospheric transparency window (8 to 12 μm), the absorption bands of different rocks are close to each other within an accuracy of up to 50%.

The report (US National Research Council 1985) recommends using the characteristics of the absorption of basalt glass, which are dependent on the wavelength (see Pollack et al. 1973).

Smoke from Nuclear Fires. To evaluate climatic consequences of a nuclear war, it is necessary to know the quantity of smoke that would be emitted into the atmosphere, how high and how far it would spread, and how long it would remain in the atmosphere. For this purpose, data on fires are required, including information on the amount of combustible materials, the fire spread, smoke emissions, optical and physical properties of smoke, the size distribution of smoke particles, etc.

It must be said that the theory of large-scale fires that would occur after the nuclear explosions has not been sufficiently developed, since it was only in 1982 that attention was drawn to the fact that, as a result of heavy fires, large quantities of smoke would be released into the atmosphere.

Historically, mankind has long been acquainted with violent urban fires. For example, in 64 A.D. at the time when Nero reigned, a fire started in Rome (population: one million) and continued for 9 days, devastating the greater part of the city, which was to a large extent built of stone. The Russian wooden towns, great and small, have periodically been completely ruined by fires. In London in September 1666, the Great Fire destroyed over 13,000 buildings within 2 days. We can also mention here the great fires in Moscow in 1812 and in Chicago in 1871. It is known that the latter was initiated by one single source of fire (Kerr 1971).

In modern times, mankind has experienced new causes of mass fires in large cities. One of them is violent earthquakes, which disrupt gas and liquid fuel lines, cause electrical short-circuits, lead to spilling out of oil fuels and so on. Particularly strong fires were observed during the earthquakes in San Francisco in 1906 and in Tokyo in 1923. It can be seen from these examples that under certain conditions urban fires spread rapidly and cover vast territories.

The fires generated by massive bombings of the cities during World War II provide some evidence for those that would accompany nuclear explosions. Thus, the fires resulting from the bombing of Hamburg on the 27th of July in 1943 burnt an area of tens of square kilometers and merged into a fire whirl, which produced smoke up to an altitude of 9 to 12 km (Ebert 1963).[2] The high altitude reached by the smoke was promoted by an almost adiabatic temperature gradient in the lower atmospheric layers. Smoke and dust covered the sky over the city for 30 h from the time of fire initiation.

As a result of two massive raids on Dresden undertaken by the British and American air forces on the 13th and 14th of February 1945, widespread fires, which continued for about a week, embraced the city. More than 75% of all the buildings within the city limits were completely burned over an area of 12 km². The fires were accompanied by a fire whirl.

Fire whirls were also observed during the bombings of Kassel and Darmstadt. However, in other similar cases, no fire whirls occurred. Subsequent investigations have shown that in order for a fire whirl to occur, a high rate of heat emission per unit area, as well as an almost adiabatic temperature gradient (10°C km^{-1}) and not very strong winds blowing at a speed of 5 to 10 m s^{-1} are necessary.

The bombings of Hiroshima and Nagasaki on the 6th and 10th of August 1945 have been the only cases of nuclear attacks against cities (during the wartime as well). In Hiroshima the central part of the city (an area of 13 km²) was destroyed and burnt to ashes within a radius of 2 km from the explosion epicenter (ground

---

[2]According to Brunswig (1982), the smoke rose up to 7 km high.

zero). A fire whirl was observed there also. The bomb at Nagasaki was more powerful, but the damage (area burnt 7 km^2) was smaller: due to a highly irregular terrain, a large portion of the city was screened against direct radiation from the thermal pulse of the nuclear burst.

All these examples show that in studying the consequences of nuclear explosions it is necessary to take account of a great number of factors: meteorological, topographic, urban settings, type of buildings etc.

According to Glasstone and Dolan (1977), during nuclear explosions at a height below 10 km, about 30–40% explosive energy goes with the thermal pulse (which lasts for about a second or several seconds for megaton bursts) of intense thermal radiation at the visible range of the spectrum (and around this spectral range), 45–55% is converted into the blast wave and about 15% is spent on the penetrating radiation and induced radioactivity. The conflagration is a direct result of the light emission (the thermal pulse), the intensity of which is usually measured in kilojoules per square meter, i.e., it is a flux of light emission integrated by pulse time. The ignition threshold strongly depends on the properties of the material irradiated, its moisture content and so on. It varies from 210 to 630–840 kJ m^{-2} (from 5 to 15–20 cal cm^{-2}). Let us indicate as an example that in Hiroshima the ignition threshold was about 300 kJ m^{-2} and in Nagasaki, 840 kJ m^{-2}. The density of emission energy that generates fires depends on the intensity of the burst and meteorological visibility in the boundary layer. Figure 15 derived from a report by the US National Research Council (1985) shows the dependence of the total density of emission energy on the distance to the explosion epicenter (in the troposphere) with different visibility. Visibility is mainly a factor of air humidity and of aerosol presence.

The thermal pulse generates the initial fires. The accompanying blast (pressure) wave may either extinguish the fire (blowing it away or covering fuels with nonflammable debris from buildings), or contribute significantly to fire growth and spread and the appearance of secondary fires by breaking up and spreading solid fuels, rupturing oil and gas lines, etc. Subsequently ambient air masses entrain the rising fireball, thus promoting the expansion of fires. There are some crude estimates (US National Research Council 1985) showing that the energy release rate during the fires at Hiroshima was appreciably lower than, say, that in Hamburg in 1943. However, in the former case, the fire developed into a firestorm, which consumed almost everything that could burn within a radius of 2 km. On the whole, there is no doubt that the fires ignited by a nuclear explosion will be more devastating than all major urban fires known in the history.

Fires can also be observed in forests and grasslands. It is interesting that the first estimates by Crutzen and Birks (1982) of smoke output were given for forest fires only. Following the data in Figure 15, the fire area produced by an explosive yield of 1 Mt may reach approximately 700 km^2. Nuclear explosions can initiate more violent forest fires than the conventional fires due to, for instance, forest tree felling by a blast wave. A rough analog for nuclear fires in the forest can be the fires initiated by the impact of the Tunguska meteorite on June 30, 1908. The intrusion of this meteorite into the atmosphere and the explosion at a height of about 8–10 km was not accompanied by the formation of a fireball (the intensity

2. Climatic Effects of a Nuclear Conflict

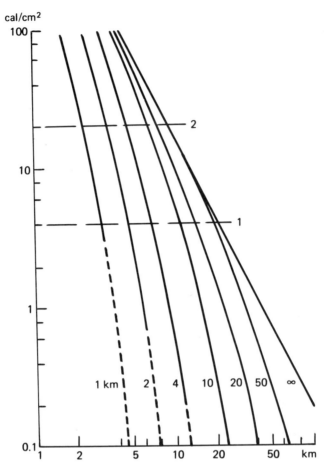

FIGURE 15. The dependence of maximum light emission on a horizontal distance at different visibilities for a 1 Mt burst at a height of several kilometers (Kerr et al. 1971). 1 = The inflammability limit in Hiroshima; 2 = the inflammability limit in Nagasaki. Light emission is approximately proportional to the burst yield.

of the light emission was a few orders of magnitude smaller than that produced by a nuclear explosion), but the blast wave, as estimated by Krinov, was identical to that which would accompany the explosion of 10 Mt of TNT at a height of about 8 km. About 2000 km² of forest fell down, many trees were barked, with branches and twigs being torn off. Numerous fires sprang up and the completeness of burning was much greater than that in a conventional forest fire in the taiga zone. However, the fire area and spread strongly depend on the season and weather conditions, the greatest fires being naturally most likely to occur in summer and in dry weather and the least severe ones in winter.

The amount of smoke released into the atmosphere resulting from fires depends on the mass and kinds of fuels and the conditions of burning. According to modern views (Crutzen et al. 1984; US National Research Council 1985; Pittock et al. 1986), urban fires would constitute the main sources of smoke.

A seminar within the framework of ENUWAR held in London in April 1984 discussed the problem of urban fires. Data were presented on fuel loads (fuel stored in cities), which showed that the fuel loads vary from 200 kg m^{-2} in the center of a modern European city to 20 kg m^{-2} in the suburbs. Using three models for cities of 1 million population, Larson and Small (1982) have estimated the total fuel stored in these towns at 10 to 40 Mt. Nowadays, there are about 200 cities with a population of 1 million people and more in the world. The Ambio scenario (1982) assumes about 1000 urban targets. Therefore the potential fuel burden constitutes approximately 10,000 Mt. This figure includes also the stored oil, gas, coal and other materials. The urban fuel load, as estimated in the report (US National Research Council 1985), is 7500 Mt, including 5000 Mt of wooden materials, 1500 Mt of oil and oil products and 1000 Mt of various plastics, synthetic fabrics and a wide variety of industrial chemicals. Pittock et al. (1986) arrived at about the same (only slightly smaller) figures; however, they made allowance of about 50% for the uncertainties in this field.

The smoke output (in percentage of the mass of the burnt material) greatly depends on the conditions of burning. The smoke mass grows rapidly with decreasing oxygen or increasing temperature of the ventilating air. Smouldering fires can produce a much greater smoke output. Thus, the smoke output from forest fires covering an area of 10 m^2 is about 3–6% by relative mass, while in the case of smouldering fires it can rise to 15%. Smoke emissions from oil products, plastics, and rubber can range from 1–15% in flames to 5–40% under smouldering conditions. From a thorough review of many sources the authors assume an average of 4%. If in the first approximation we assume that half of the potential fuel stores is burnt, the smoke output will be 200 Mt.

On the basis of detailed calculations the authors of the report (US National Research Council, 1985) estimate the urban smoke at 150 Mt. In this calculation they assume that the limit of the energy load starting the conflagration approaches maximum (840 kJ m^{-2}), that with multiple explosions one third of the inflicted areas overlap each other and the fires do not spread. Making allowance for possible uncertainties, the smoke mass is estimated at 20–450 Mt. The limit of 20 Mt has been obtained with a 1% smoke output and the urban fuel burden of 20 kg m^{-2} as compared to 40 kg m^{-2} in a baseline case.

Let us now discuss forest fires. According to Safronov and Vakurov (1981) the dry wood store in forests is on the average about 15 kg m^{-2}. It can range from 1 kg m^{-2} to 30 kg m^{-2}, reaching 25–30 kg m^{-2} in highly productive forests, where 15–20% of dry biomass (3.5–6 kg m^{-2}) constitute flammable litter, or kindling (dry twigs, leaves, etc.), which as a rule, burns completely. Out of the biomass of forest trees, about 20% (twigs that are up to 4 cm thick, bark and parts of tree trunks) are usually burnt completely. On average, approximately one third of the dry forest biomass is burnt, which at a dry wood store of 15 kg m^{-2} constitutes about 5 kg of burnt material per one square meter of forested area.

The fires in peat, where the burnable material can reach from 0.5 to 15 kg m^{-2} (Safronov and Vakurov 1981), have not yet been considered adequately. The summer fires in peat under smouldering conditions can last, as in 1972 in the USSR, up to the autumn rainy season.

In experiments with a stack of wood (Devlishev et al. 1979), the smoke output, according to lidar measurements, was about 2%. According to Crutzen et al. (1984), the burning of different kinds of trees yields a smoke output from 0.5 to 7%, with the most probable value of 4%. In the report (US National Research Council 1985) it is assumed that the fire consumes 4 kg of wood per m² of forest area (the figure is similar to that of Crutzen et al. 1984), but the smoke emission is 3%. Thus, it can be seen that independent works are in agreement in estimating both the "fuel" density and the smoke output. The latter is estimated as 0.1 kg m^{-2} by the Soviet scientists, 0.16 kg m^{-2} by Crutzen et al. (1984) and 0.12 kg m^{-2} by the authors of the report (US National Research Council 1985).

The first estimates by Crutzen and Birks suggested that the fire can consume about 10^6 km² of forest (10^{12} m²). In this case the smoke output will be 100 to 160 Mt (in the initial calculations of these authors the smoke emission was 200 Mt, assuming that 5 kg m^{-2} of wood is burnt with the smoke output of 4%). In the baseline scenario (US National Research Council 1985), it is assumed that the forest area consumed by the fire is 2.5×10^5 km², which yields 30 Mt of smoke. Allowance is also made for the possibility of generating 200 Mt of smoke by forest fires covering an area of 10^6 km² when the smoke output is 5%. The report takes the absence of smoke from forest fires as the lower limit in these calculations of the total smoke amount.

In Chapter 1, when considering forest fires, we have already mentioned that under ordinary conditions smoke reaches a height of 2 to 3 km, although there have been cases when it rose much higher. At first, smoke rises in the so-called thermals, i.e., the air heated by the fire. The thermals reach the height at which the air density within them approaches the density of the ambient air. The rising air in thermals can expand and cool at the same time. The thermals can also entrain the ambient air.

The height to which smoke rises and the possibility of the upward convective air flows from fires entering the stratosphere can be roughly estimated by the theory of turbulent jets. In the case under consideration, the calculated results can be found in Izrael's monograph (1984).

The numerical calculations for a two-layer atmosphere were performed by Wolfson and Levin (1981). It is shown in this work that with a heat source of 10^6 kW (which corresponds to 70 t of kerosene burnt per hour), in a dry standard atmosphere the plume rises to a height of about 2000 m. It has been found that the height reached by the plume is proportional to the fourth root of the heat power of the source (see also Gostintsev et al. 1985; Manins 1985).

In calculating the plume rise to a great height it was assumed that the tropopause was 11,000 m high with stratification as a standard atmosphere ($\gamma = dT/dz = -0.65°C/100$ m) and the stratosphere was characterized by isothermal conditions ($\gamma = 0$). As boundary conditions near the Earth's surface, it was assumed that the initial radius of the plume R_0 was 500 m, the vertical air speed W_0 was 20 m/s, and the temperature within the plume exceeded the ambient air temperature by 100, 200, and 300°C (ΔT_o). These conditions corresponded to the heat source $Q = \pi R^2_o \rho c_p W_o \Delta T$ (ρ is the air density, c_p the air heat capacity at a constant pressure) of about 0.9×10^9 kW, 1.85×10^9 kW and 2.8×10^9 kW.

These calculations have shown that for the maximum heat source the peak altitude reached by the plume is 12 km and more.

Thus, with a heat source of $(1 \div 2) \times 10^9$ kW, the plume can enter the stratosphere. Such a heat power theoretically corresponds to a forest fire covering an area of 10 km² when the burning rate is 3–5 g(m² s⁻¹), i.e., with a complete burning of all the wood in the forest within 2 to 3 h (which is never realized in practice).

The phase transformations produced by the presence of moisture in the atmosphere, the quantity of which increases with the wood burning, facilitate the process of the air rising in convective flows due to the heat release by condensation.

Calculations have shown that the vertical air velocity in the plumes rising in humid atmosphere is approximately twice that in dry atmosphere. The top of the plume from the meteotron was about 6000 m in standard, moisture-saturated atmosphere and 2000 m in dry atmosphere with the same source power.

The information necessary for evaluating the smoke rise in thermals of different types is summarized in the work by Manins (1985).

Under conditions when it is possible to neglect the effects of wind (more accurately, the wind direction shear with height) and with an average rate of temperature decrease with height of about 6°C km⁻¹, the height reached by the top of the thermal is estimated as

$$Z_{u.b.} \approx 0.25 \, Q^{¼},$$

where Z is the height reached by the thermal in km and Q is the heat source power in MW. Figure 16 (after Manins 1985) is supplemented with a number of dots. It shows the altitudes of the thermal rise from 20 volcanic eruptions and several massive forest and urban fires. The solid line corresponds to $Z_{u.b.}$ given in the above formula. The height of the lower boundary of an isolated thermal (if the heat source is of shorter duration than the thermal takes to rise) is estimated as $Z_{l.b.} \approx 0.67 \, Z_{u.b.}$. To reach the mid-latitude tropopause (at 10 to 11 km), the heat source power must be 2.4×10^6 MW (which agrees with the previous analysis); to inject half the smoke into the stratosphere, the heat power must approach 10^7 MW.

To check this formula, let us apply it to the results of the numerical simulation by Cotton (1985) for a hypothetical large city fire. The experiment was made for a burning area of 8 km in diameter with a heat output of 10^5 W m⁻², which totals 5×10^6 MW (about three times that of the fire in Hamburg in the summer of 1943). The above formula gives $Z_{u.b.} = 12.1$ km. According to Cotton, the top altitude is 14 km, and 44% of smoke is injected into the lower stratosphere. For Hamburg, the same formula yields $Z_{u.b.} = 9$ km, which is also in good agreement with the observed fires, where smoke rose to 7–12 km (Brunswig 1982; Ebert 1963). Similar results have also been obtained in numerical simulations of massive fires carried out in the Lawrence Livermore

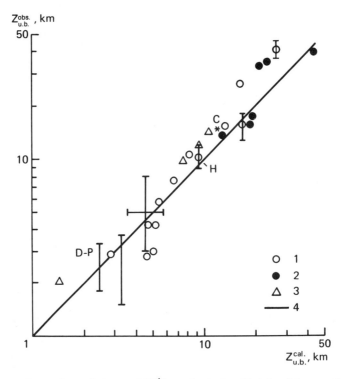

FIGURE 16. Comparison of observed ($Z_{u.b.}^{obs}$ and calculated height of the top of the cloud formed as a result of major fires and volcanic eruptions. (After Manins 1985.) 1 = Middle latitudes; 2 = the tropics; H = clouds over Hamburg in 1943; $D\text{-}P$ = the results of observations of a fire in a petroleum plant (Davies 1959); C = the results of a numerical experiment by Cotton (1985); 3 = the calculated results by Wolfson and Levin (1981); $4 = Z_t = 0.255\, Q^{1/4}$ (Q is the power, MW); $Z_{u.b.}^{obs}$ is the observed height of the upper boundary of the thermic; $Z_{u.b.}^{cal}$ is the calculated height of the upper boundary of the thermal.

National Laboratory, USA (Penner et al. 1985; Penner et al. 1985, and also Pittock et al. 1986).

Since the intensity of the fires is different, it can be assumed that the smoke distribution will be uniform with altitude between 1 and 10 km according to Turco et al. (1983) or up to 9 km according to the report (US National Research Council 1985). It is, however, quite possible that a considerable mass of smoke will also be injected into the lower stratosphere.

Taking all these considerations into account, it is unlikely, with a few exceptions for low atmospheric stability, that considerable quantities of aerosol particles from forest fires can be directly injected into the stratosphere. Injection of smoke mass into the stratosphere is more probable during urban fires.

One further mechanism (in addition to the increased air humidity, which has already been mentioned) can promote smoke rise. As a result of the absorption of solar radiation by smoke, the atmosphere permeated with smoke will be heated as compared to the upper adjacent clear air layers. This must lead to an additional buoyancy of the smoke-laden air. Ginzburg et al. (1985) obtained a rough estimate of the efficiency of this mechanism, by introducing into the model describing the growth of the diurnal convective boundary layer, the volume absorption of solar radiation (uniform with height). Then, the increasing thickness h can be estimated as a factor of time t by the following formula:

$$h(t) = (2Qt/\Delta\gamma\rho c_p)^{1/2},$$

where Q is about 120 W/m^{-2}, representing an average diurnal positive radiation balance at the top boundary of the absorbing layer; $\Delta\gamma = \gamma-a \approx 3.5°C$ km^{-1}, the vertical temperature gradient deviation from the adiabatic one ($-9.8°C$ km^{-1}); ρc_p, the enthalpy of the unit of air volume. The results obtained show that for approximately 10 days this mechanism can raise the top boundary of the smoke-laden air layer from, say, 2 to 10 km, i.e., up to the lower boundary of the stratosphere.

The mechanism of the ascending heated smoke-laden air has been well reproduced in multi-layered models of the general circulation, which also include the stratosphere. These models have been used for simulation of the worldwide consequences of nuclear war in the National Center for Atmospheric Research in Boulder, USA (Thompson 1985) and in the National Laboratory in Los Alamos, USA (Malone et al. 1986); the model results have also been described by Pittock et al. (1986). These models and others have shown that considerable quantities of smoke can be injected into the lower stratosphere as high as 20 or 30 km.

Turco et al. (1983) have estimated that urban fires associated with greater fuel burdens (on the average about 3 g cm^{-2} and in city centers 10 g cm^{-2}) produce 52% of the total amount of aerosol particles, whereas firestorms generate 7% of aerosol and forest fires (with an average fuel load of 2.2% g cm^{-2}), 34% in the first 10 days and 7% in the subsequent month. Urban fires consume 1.9 g cm^{-2} of flammable materials, and forest fires 0.5 g cm^{-2} (over an area of 5×10^5 km^2). The total amount of smoke will then be 2.25×10^8 tons, of which 5% (or about 11 Mt) reach the stratosphere. This will lead to a rapid increase in the optical density of the atmosphere.

One important problem appears to be the estimation of the lifetime of smoke particles. Under ordinary conditions with a small amount of smoke, when it is possible to neglect its effects on the thermal regime of the atmosphere, the behavior of smoke is similar to that of the conventional aerosol. In this case its lifetime, τ, can be estimated by the data given in Figure 14. For the lower troposphere it will then be several days or a week, for the middle troposphere from 10 to 15 days, and for the upper troposphere about a month. In the lower stratosphere, for submicron smoke particles τ can already be many months.

It is important to know for how long smoke is being emitted into the atmosphere. It has been estimated that during times of peace the total annual release of smoke into the atmosphere all over the Earth is 200 Mt (Seiler and Crutzen 1980), which is close to the estimates for smoke injections from fires during a nuclear war. Under normal conditions the major sources of smoke are the combustion of fossil fuels (oil, coal, natural gas), forest and other types of fires, for instance, those associated with land cultivation. The "peaceful" smoke differs from that of the wartime, and the conditions of its injection into the atmosphere are also different. Its most prominent feature is the low burning rate, as a result of which the greatest smoke mass is concentrated in the atmospheric boundary layer, i.e., in the lower layer about 1 km thick. From here, the smoke particles are quickly removed by precipitation. Moreover, the injection of smoke occurs in different places and more or less uniformly throughout the year, therefore it nowhere accumulates in quantities that can appreciably influence the thermal regime of the atmosphere. Additionally, the elementary carbon content of the smoke particles is not high, since the greatest mass of smoke is generated through the combustion of wood under controllable conditions. The average lifetime of smoke particles averaged throughout the atmosphere is 10 days or less (Ogren 1982).

Hence, it can be estimated that at any given moment there are about 5.5 Mt or less of smoke in the atmosphere. These smoke particles are not highly absorbent for solar radiation. As a result, the background concentrations of elementary carbon are usually 0.1 µg/m^3 or less, and its optical density by absorption is of the order of 0.01 (Charlson and Ogren 1982).

The published nuclear war scenarios (Ambio 1982; US National Research Council 1985; Pittock et al. 1986) proceed from the assumption that a major nuclear exchange will be a matter of only a few days. Based on experience, it is suggested that urban fires will continue for 24 h with the highest intensity during the first few hours, while forest fires will last about a week. At such an injection rate, the atmosphere will accumulate a considerable smoke mass even with a normal removal rate.

However, there are grounds to think that the lifetime of smoke particles can be prolonged after large quantities of "wartime" smoke are introduced into the atmosphere. As has already been mentioned in Chapter 1, the most effective process for removing smoke particles is scavenging by precipitation. Under normal conditions, most precipitation is formed in the lower part of the troposphere. Therefore an injection of large quantities of smoke (of the order of 50% or more of its mass into the upper troposphere ($Z > 5$ km) or even lower troposphere appreciably increases the average lifetime of the smoke particles, which is usually from 5 to 10 days. Moreover, it can be expected that the intensity of the moisture cycle will decrease (see more details below), which will delay the removal of the smoke by precipitation, simply because precipitation will become lighter and will be formed in lower layers of the troposphere. These simple qualitative aspects of the problem raised by Obukhov and Golitsyn (1983, 1984) have been confirmed in numerical experiments by Malone et al. (1986), where all these processes are described in detail.

Let us mention one more cause for the decrease in precipitation: a fast increase in concentration of aerosol particles in the air after the injection of smoke into the atmosphere will increase the number of condensation centers, i.e., at the same relative humidity (which will actually be decreasing in the heated atmosphere) the growth rate of water droplets will be delayed because of their competition (overseeding). Therefore, the droplets may not reach a size sufficient to fall as precipitation (Crutzen and Birks 1982; Obukhov and Golitsyn 1983, 1984).

Let us consider the properties of the smoke and how they change with time. Figure 14 shows the typical size distribution of forest-fire smoke particles (curves 4 and 5). This distribution has been obtained by analyzing the air samples taken by Stith et al. (1981) from aircraft at different levels and distances from a large burning pile of forest litter. It can be seen from the figure that the distribution has a major peak at about 0.1 μm and 0.3 μm, i.e., in the size range of the particles with the longest residence time. The particle size determines both the lifetime of the particles and the radiation transport in the smoke-laden air layer. The size distribution depends on the properties of the flammable material and the conditions of burning. It approaches the logarithmically normal distribution with dimensionless semi-width of $\sigma = 2$. For such distributions, the mass median particle radius r_m (i.e., the radius dividing the size distribution into two equal parts by mass) is 4.3 times greater than the number modal radius that coincides with the peak of the size distribution of numbers of particles. For timber, r_m is close to 0.2 μm for flame conditions and up to 0.6 μm under smouldering, the soot content (or elementary carbon) being taken as 8 to 30%. For different synthetic materials $r_m = 0.6$ μm. The burning of plastics and synthetic rubber produces about 90% and more soot of the mass burnt. As can be seen, the smoke generated in burning of these materials contains particles in the size range of the so-called "Greenfield gap" (see Fig. 14), i.e., the particles with the longest residence time in the atmosphere. Moreover, the soot smoke is hydrophobic, at least during the first day of its existence.

Coagulation that occurs by the Brown diffusion is very important for soot particles, mainly because it leads to an increase in particle size. Therefore, the total particle number in a smoke cloud does not exceed 10^5 cm^{-3} in a few hours and 10^4 cm^{-3} in about a week (Twomey 1977). Coagulation of smoke particles from nuclear fires has been considered in models of the atmospheric consequences of a full-scale nuclear exchange (Turco et al. 1983; Crutzen et al. 1984). If the initial smoke concentration close to the intense fire is of the order of 10^6 cm^{-3} for urban and controllable fires and of the order of 10^5 cm^{-3} for forest fires, a further expansion of the upward air flows decreases the smoke concentration, the process of coagulation is strongly retarded (since its rate is proportional to the square of the particle concentration), and usually in about half an hour the particle size distribution becomes stabilized and subsequently does not appreciably change. Reviewing the data on controllable and forest fires led the authors of the report (US National Research Council 1985) to consider the number modal radius of 0.1 μm to be characteristic of the majority of the smoke particles, independent of their origin and age (not considering the first minutes or the first hour of their lifetime).

The air in rising thermals expands and becomes cooled, which may lead to condensation of both the water vapor entrained from the ambient air and the water vapor produced by burning. These droplets can attract soot particles and then fall as a black rain similar to that observed in Nagasaki and Hamburg, a process determined by meteorological conditions and the intensity of the fire. According to Knox (1985), this problem has been investigated by numerical models in the Lawrence Livermore National Laboratory. It has been found that, depending on meteorological conditions, the black rain can scavenge up to 25% of the particles introduced into the atmosphere (in the case of Nagasaki, about 3% of the particles were washed out in this way, which agrees with some of the indirect estimates (Molenkamp 1979).

If the rain does not fall immediately, the water droplets can subsequently vaporize, and the smoke particles entrained into the water droplets tend to agglomerate into one large particle. This process can also influence the size distribution of the particles and their concentration.

The report of 1985 describes such a process as was observed by Radke, who studied smoke transformation in a cloud formed above a forest fire at a height of about 2 km. He collected the samples of smoke particles from aircraft both below and above the cloud (i.e., the smoke that was "filtered" through the droplets in the cloud). In the smoke above the cloud the number of particles with radius $r \leq 0.05$ µm was almost an order of magnitude smaller than below the cloud. The number of particles with radius $r \geq 0.5$ µm also decreased several times, while the number of intermediate particles practically did not change, although the number modal radius increased from 0.05 to 0.1 µm. However, the total mass of submicron particles hardly changed and the extinction coefficient within the limits of 20% remained the same. On the whole, the droplets in the cloud can influence the distribution and concentration of the smoke particles. However, the few available data have shown that this influence is not very strong. Moreover, simple physical considerations suggest that cloudiness in the smoke-laden and heated atmosphere cannot be very heavy, and calculations by the general circulation models, taking account of cloud formation and changes in the thermal regime because of the absorption of solar radiation by smoke, show that cloudiness sharply decreases in the smoke-laden atmospheric layers (Malone et al. 1986) as well as precipitation. On the whole, the process of transformation and evolution of the smoke from fires is still largely unknown (Pittock et al. 1986).

The chemical composition of the aerosol determines its complex index of refraction. The real part of this index characterizes the aerosol scattering properties, and the imaginary one the absorbing properties. If particles have a complex or multilayer structure, the refraction index can be determined only by measurements. Turco et al. (1983) have taken m = $1.5-0.001i$ for dust and $1.7-0.3i$ for urban-fire smoke, which corresponds to a high soot concentration, while Stith et al. (1981) have obtained m = $1.53-0.05i$ for forest-fire smoke. In the report (US National Research Council 1985) in the baseline case for smoke, m is taken on average as $1.55-0.1i$, the minimal value of $1.5-0.002i$ to that used by Turco et al. (1983).

To calculate the transmission of both solar and thermal radiative, the radiative transfer equation is used in one or another approximation (see Liou

1980). It is usually sufficient to have the so-called transport and two-stream approximations (the delta-Eddington approximation). Generally speaking, the calculations should be carried out for spectral intervals if the absorption and scattering depend on the wavelength λ. Such a dependence always exists, since the Mie parameter $\rho = \pi d/\lambda$ (where d = $2r$ is the diameter of a spherical particle) determines scattering and absorption. Because the modal radius of the particle size distribution approaches 0.1 μm, the Mie parameter is close to a unit in the maximum solar intensity of the spectrum $\lambda = 0.55$ μm where sunlight interacts most intensity with the particles. However, since there is a particle size distribution, the effects are smoothed over the wavelength of the spectrum, and it is possible to introduce the average absorption and scattering coefficients for the entire solar spectrum and also for thermal radiation (Pittock et al. 1986).

The optical properties of the aerosol layer are characterized by its optical thickness, τ_e, so that the intensity of direct radiation flux incident on the layer at an angle of θ is attenuated by $\exp(-\tau_e/\mu)$, where μ is the cosine of the zenith angle. $\tau_e = \sigma_e M$, where M is the mass (g m^{-2}) of the absorbing aerosol in a vertical column of the layer (with the base area of 1 m^2) and σ_e is the extinction coefficient equal to the sum of the coefficients of the absorption (σ_a) and scattering (σ_s). The attenuation of the direct radiation flux incident on the aerosol layer is composed not only of the absorption and scattering in the forward hemisphere (where at multiple scattering, radiation might be additionally absorbed) but also of the scattering in the backward hemisphere. At the same time, a portion of radiation is reflected from the layer. Scattering and reflection are determined by the scattering indicatrix (phase function), which depends on particle size and refraction index. The greater the imaginary part of the complex index of refraction is, the less radiation is scattered and reflected. The dependence of sunlight transmission on the optical thickness of the smoke and dust layer is presented in Figure 8.

In a thermal range with a typical wavelength of $\lambda = 10$ μm, the Mie parameter for smoke particles is approximately 1/20; therefore, smoke exerts little influence on thermal radiation. Considering the size distribution of particles and their optical properties, a report by the US National Research Council (1985) recommends the following basic parameters: the sunlight extinction coefficient for smoke 5.5 (2 ÷ 9) m^2/g (in parentheses are given the lower and upper limits of possible variations of the value), the absorption coefficient in the visible region 2 (1 ÷ 6) m^2 g^{-1}, the absorption coefficient in the thermal region 0.5 (0.2 ÷ 5) m^2 g^{-1} (in this region the scattering for particles with the Mie parameter 0.05 is negligibly small). These numbers show that in the thermal region the smoke layer affects the radiation flux by about an order of magnitude weaker than that in the visible range.

An important factor in calculating the radiation transfer is the so-called single-scattering albedo, defined as $\omega = \sigma_s(\sigma_s + \sigma_a)^{-1} = \sigma_s/\sigma_e$, i.e., the ratio of the scattering coefficient to the attenuation (extinction) coefficient. For the adopted magnitudes of these coefficients, $\omega = (5.5-2)/5.5 = 0.64$. In reality, $\omega = 0.5$ for smoke with high soot content and $\omega = 0.8-0.9$ for forest-fire smoke (Crutzen et al. 1984). Kondratjev et al. (1984) review the ω values for different aerosols. Depending on the soot content, ω can vary from 0.17 to 0.94. It is obvious that ω can somehow change in the process of aerosol evolution.

Concluding this section, we give estimates of the optical density of a smoke cloud of 150 Mt in mass, if it is uniformly distributed over (1) a region of 30° to 70°N lat. (44% of the hemisphere's area) and (2) the entire Northern Hemisphere. In the first case the columnar smoke mass $M = 1.34$ g m^{-2} and with $\sigma_e = 5.5$ m^2 g^{-1}(±3.5) we obtain $\tau = 7.4$ (± 4.7). In the second case $\tau_e = 3.2$ (± 2.0). According to Figure 8, in the first case the direct sunlight is attenuated more than a thousand times, and in the second one by a factor of about 40.

The sunlight absorption by the smoke layer should lead to the warming of the atmosphere, while a strong attenuation of the sunlight intensity near the surface significantly changes the energy balance at the surface. All this should alter the thermal and, consequently, dynamic regime of the atmosphere and the atmosphere's interaction with the underlying surface. Now we shall describe these changes predicted by models of different complexity and their natural analogs, although the latter may not be perfect.

The smoke rising from an individual fire (not considering subsequent heating by the Sun) generally remains at the level of its buoyancy, i.e., at the height where the density of the smoke-laden air is equal to that of the pure air. Far from the smoke source, the near-surface air might be quite pure, as was the case, for instance, in the eastern part of the United States at the end of September 1950, when the sky became turbid with smoke coming from fires in Western Canada. According to the estimates by Smith (1950) and Wexler (1950), the smoke spread at a height of about 2 km and higher. The layers of the Arctic haze are also usually found at a height of 2 to 5 km. At these levels the aerosol is carried around with the wind.

The free atmosphere (the layers above the boundary layer of about 1 km thick) is practically always stably stratified. Therefore the turbulence there is strongly attenuated and occurs only in patches (which can be observed in flight, when the rough air is felt only at certain moments). This weakens the vertical mixing of aerosols injected into the free atmosphere as compared to their injection at the surface, i.e., into a well-mixed atmospheric boundary layer where convection is developed at least during the daytime. In some studies simulating atmospheric consequences of a nuclear war (Turco et al. 1983; Covey et al. 1984) it was therefore assumed that the smoke density (by mass) is uniformly distributed by altitude, which is contrary to the assumption of constant aerosol ratio when the aerosol density decreases with height with decreasing air density. This reflects to a certain degree the variability in height of many smoke plumes from different fires.

The dust storms on Mars are an example of the optically active dust that spreads rapidly in the atmosphere by the arising wind systems of different scales. Qualitatively, it is possible to expect the same mechanism to operate in the Earth's atmosphere, where although the air density is higher, the absorbing properties of the smoke are stronger than those of Martian dust.

The description of smoke spread in the atmosphere is given in the general circulation models, where smoke is injected in certain regions of the Earth (in some continental mid-latitude regions of the Northern Hemisphere) and carried away by the wind; it heats the atmosphere and absorbs solar radiation (Stenchikov 1985a; MacCracken and Walton 1984; Covey et al. 1984; Malone et al. 1986; Pittock et al. 1986). All these experiments show that the heated smoke-laden air rises into the upper layers of the simulated atmosphere and first spreads horizontally in the zonal direction, covering within about a week the entire mid-latitude region, including the oceans. Simultaneously, the smoke is being transferred to the tropical areas and further to the Southern Hemisphere, into which it penetrates in 2 or 3 weeks after its appearance in the northern atmosphere.

Although the modern general circulation models do not yet consider the meso-scale processes, they nevertheless show that smoke spreads globally rather rapidly. Consideration of meso-scale factors and the feedbacks between smoke (i.e., the resultant heating of the atmosphere) and wind field can only accelerate the mixing processes in the atmosphere. Smoke distribution also depends on the season. In winter when the Sun's altitude is not high, the smoke reaches lower altitudes and spreads less intensely into other latitudes than in summer.

In these general circulation models, smoke scavenging is considered only very roughly (sometimes simply as a decrease in smoke concentration with time). It may well be that to obtain more reliable estimates of nuclear war consequences in the tropics and in the Southern Hemisphere it will be necessary to apply sufficiently detailed physical schemes for removing the smoke from the atmosphere.

THE ENERGY BALANCE IN THE ATMOSPHERE

Before turning to estimating the influence of the smoke cloud on the temperature regime, let us briefly show how this regime is established under standard conditions by different heat inflows to the atmosphere and the Earth's surface.

The thermal regime of the Earth is determined by the solar radiation energy coming to the Earth and the energy transformations in the atmosphere, thanks to which the atmosphere and the Earth's surface are heated and emit the energy into space in the form of thermal radiation. A planet with the radius a receives the solar radiation flux with the intensity q equal to 1.37×10^3 W m^{-2} at an average distance from the Earth to the Sun. A portion of this energy, determined by the reflecting coefficient, or the planetary albedo A, is reflected and scattered into space. The value A is determined by the reflecting, absorptive and scattering properties of the atmosphere and the underlying surface, and also by the presence in the atmosphere of aerosol layers or clouds (including those containing water).

The overall amount of solar radiation coming to the atmosphere and the planet's surface is $\pi a^2 q(1-A)$, where πa^2 is the area of the circle on which the solar radiation flux is incident. Thermal radiation is emitted by the entire area of the planet, $4\pi a^2$, with intensity equal to $4\pi a^2 \sigma T_e^4$, according to the law of Stefan-Bolzmann, where σ equal to 5.67×10^{-8} W/(m^2 K^4) is the constant of emission, and T_e is the effective temperature of outgoing radiation. On the average, the incoming solar radiation flux must be counterbalanced by the flux of outgoing thermal emission of the planet, because the planet as a whole is on average

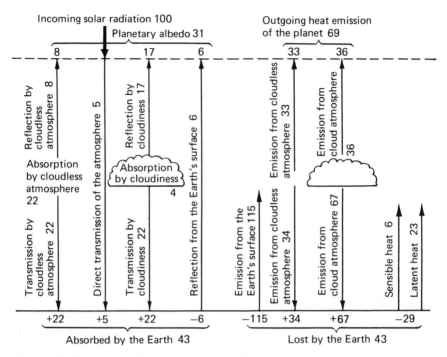

FIGURE 17. Conversion of solar and thermal radiation in the Earth's atmosphere. Figures are given as a percentage of the mean solar constant.

neither heated nor cooled under standard conditions. Making these two fluxes equal to each other, we obtain the equation for determining the outgoing radiation temperature:

$$T_e = [q(1-A)/4\sigma]^{1/4}.$$

The value T_e characterizes the planet's temperature regime. For the Earth, according to numerous satellite data, $A = 0.3$ (i.e., about 30% of the incoming solar energy is reflected back to space) and then $T_e = 255$ K $= -18°C$.

However, the area and year average surface temperature of the Earth is equal to about $+14°C = 287$ K, i.e., it exceeds T_e by 32°C. This difference is explained by the greenhouse effect of the atmosphere, which means that the atmosphere is more transparent to solar radiation than to thermal radiation.

The consecutive transformations of energy in the atmosphere are given in Figure 17. Some of the numbers taken from different sources (for instance, Budyko 1971, 1980; US National Academy of Sciences 1975; Liou 1980; Bolle 1982) can differ from each other by 10 to 15%, which shows that our knowledge of the Earth's energy regime is imperfect.

If we confine ourselves to considering only the solar and thermal radiation fluxes at the upper and lower boundaries of the atmosphere, we can introduce the so-called transmission functions separately for the two types of radiation (Goody 1964; Feigelson 1970; Feigelson et al. 1981). They show the amount of radiation that comes from one boundary of the atmosphere to the other. For solar radiation the portion is $D_s \approx 0.5$, while for thermal (D_T) radiation it is 0.14. In the Earth's atmosphere thermal radiation is mainly absorbed by water vapor, the major mass of which is concentrated in the lower 3-km layers, and to a lesser extent by carbon dioxide and even less by ozone, methane, nitrous oxide, and other gases. The level where outgoing thermal radiation is formed is found close to the height at which the optical depth in the thermal region τ_T is about unity. Under standard conditions this level occurs at about 5 to 6 km (i.e., not at the surface, as would have been the case without the greenhouse effect).

The Earth's surface temperature is determined by the conditions of its energy balance. As can be seen in Figure 17, the surface receives solar radiation and longwave radiation emitted by the atmosphere. Loss of energy occurs as thermal emission of the surface, heat flux into soils, actual and latent heat fluxes, i.e., heat expenditure on water evaporation from the surface of land and oceans. The latter two fluxes are of turbulent nature, and depend strongly on the dynamic and thermal state of the atmosphere. With unstable stratification of the atmosphere, when the air close to the surface is lighter than in the overlying layers, these fluxes are directed from the Earth's surface to the atmosphere. With stable stratification, these fluxes are directed from the atmosphere to the Earth's surface. Their absolute values greatly depend on wind velocity, but on the whole these fluxes are much less intense (in absolute values) for stable lower air layers than for unstable ones (i.e., in the presence of convection).

The relative values of all the components of the energy balance equation at the surface change greatly according to day, night, or season, depend on weather, and vary from place to place. For our purpose it is useful to draw an analogy with the temperature course through day and night or temperature variations during the polar night. On land, at sunset, the soil temperature starts dropping because of thermal radiation emission from the surface. With clear skies the cooling is comparatively quick, but with cloudladen skies, this process is much slower, particularly with low clouds that absorb the thermal radiation and emit part of this radiation back at almost the same temperature. At night the soil surface usually becomes cooler than the air, and in the air the night temperature inversion develops within the air layer several hundred meters thick. In winter or at the polar night, the inversion might occur in a layer of up to 1 km thick and more, with the temperature within this layer increasing by 15° to 20°C as compared to the surface temperature.

Under such conditions of strong hydrostatic stability, the wind velocity is usually not high and the turbulence is of a purely mechanical origin, its intensity being considerably attenuated by the stable stratification.

In the absence of solar radiation, evaporation decreases greatly. The heat flux into the soil does not as a rule exceed 10% of the remaining components.

Above the oceans the conditions of thermal balance differ considerably from those above the land surface. Because of high thermal inertia, the surface ocean layers practically do not cool down during the night. Overnight, the free atmosphere has time to cool as compared to the ocean. Therefore, at night above the ocean surface unstable stratification develops, that in the tropics leads to the formation of small cumulus clouds by the morning. A similar change in the thermal regime of the ocean occurs in winter in middle latitudes, when the temperature of the upper water layer is often much higher than that of the lower air layers. Let us consider the question of how the air temperature is affected by the energy of explosions and subsequent fires. In order to estimate these effects, it is necessary to know the heat capacity of the atmosphere. For a vertical air column with a base of 1 m^2, this value is 10^7 J/(m^2 K) and for the entire atmosphere of the Northern Hemisphere is comprises 2.5×10^{21} J/K. The total energy produced by an explosive yield of 5000 Mt is 2.1×10^{19} J. Comparing these figures, we find that the temperature increase averaged over the entire atmosphere is about 0.01 K. As has already been mentioned, fires provoked by nuclear explosions can consume up to 10^4 Mt = 10^{13} kg of fuel with heat production of $\sim 2 \times 10^7$ J/kg, i.e., a total of about 2×10^{20} J of energy will be produced, which is higher, by one order of magnitude than the direct production of energy by explosions. However, even that amount of energy can heat the atmosphere on the average by no more than 0.1°C. Of course, close to the sites of the explosions and fires, the local heating of the atmosphere can be great.

All given estimates of energy release show that this is small compared to that received by the Earth from the Sun. For one hemisphere, this source of energy daily yields about 6×10^{23} J, i.e., an energy amount 3000 times greater than that from possible fires. As a result of such a comparison, it can easily be understood why the attenuation of this source of energy can drastically affect the temperature of the Earth's surface and of the lower atmosphere.

Besides the studies described in Chapter 1 and at the beginning of Chapter 2, since 1982 a great number of studies have been devoted to estimating the effects of aerosols, namely of dust and smoke clouds that result from a nuclear conflict. These models are elaborated to a different extent with different spatial and temporal resolution of the processes. We have already mentioned a radiative-convective model used by Turco et al. (1983) to evaluate the average hemispheric effects, as well as the general circulation models studying the evolution of the processes in space and time over the entire globe for a period of one month or longer (Alexandrov and Stenchikov 1983, 1984; Covey et al. 1984; MacCracken and Walton 1984; Thompson 1985; Malone et al. 1986). Even though these models describe the atmospheric processes to a highly different extent of elaboration they investigate one simple mechanism: if aerosols absorb the greater portion of solar radiation for a sufficiently long time period, the lower atmospheric layers become drastically cooled.

The simplest theory of this phenomenon has been developed by Golitsyn and Ginzburg (1983, 1985). Their model takes into account the balance of only radiative types of energy at the upper boundary of the atmosphere, in the atmosphere itself, which is considered as a single layer averaged over height, and at the Earth's surface. This model is convenient, because it produces quite simple analytical formulas to evaluate the average effects of temperature changes at the surface and in the atmosphere. The calculated results agree satisfactorily with the observed temperature changes during dust storms on Mars. They also describe reasonably well the modern average climate on the Earth and Mars, and agree with the estimated temperature changes at the time of the probable asteroid catastrophe 65 million years ago (Toon et al. 1982; Pollack et al. 1983). There is thus the hope that such a model should yield reliable estimates also for large smoke clouds, at least for average conditions in the interior of the continents and above the oceans.

We give here the basic principles of this theory and its consequences. A smoke cloud absorbing the solar radiation effectively prevents its access to the Earth's surface. The radiation is mainly absorbed (63%) in its upper layers up to $\tau = 1$, which is also where the atmosphere is heated most. The surface of land and subsequently the lower atmospheric layers start to cool down more quickly under such conditions. Although at the same time the infrared opacity of the atmosphere is also somewhat increased, the thermal radiation of the Earth and the lower atmospheric layers leads only to a slower cooling than under clear night conditions. Calculations based on three-dimensional general circulation models taking all smoke properties into consideration have shown that even within a few days the temperature in the interior of the continents falls below 0°C (see Pittock et al. 1986).

In this case, the greenhouse effect has practically ceased to influence the climate. In models with high resolution in altitude, this occurs because the smoke occupies almost the whole troposphere and solar radiation absorption occurs above the layer where the major fraction of water vapor, the basic factor sustaining the greenhouse effect in the atmosphere, is concentrated. In the simple models, such as that of Golitsyn and Ginzburg (1983, 1985), this occurs because the atmosphere begins to be less transparent for solar radiation than for thermal.

Simple considerations also allow us to evaluate the asymptotic value of land surface temperature after the solar energy ceases to reach the Earth's surface. The only energy source is then the thermal radiation of the atmosphere, i.e., the atmospheric layers heated at the expense of solar radiation absorbed by the smoke layer. Since this absorption takes place in the layers where the amount of water vapor is small and the greenhouse effect does not operate, these layers can be heated only up to the effective temperature, T_e. If in the first approximation we assume that, with the appearance of the aerosol cloud, the albedo of the Earth-atmosphere system changes comparatively little, then under radiative equilibrium the surface temperature for the optically thick smoke cloud should be $T_e \approx 255°K = -18°C$. These simple considerations are confirmed by detailed calculations by Pollack et al. (1983), who simulated changes in the Earth's surface temperature with the help of a radiative-convective model that took into account the radiation interaction with the dust raised by an asteroid. They are also confirmed by calculations carried out by Golitsyn and Ginzburg (1983, 1985).

Figures 18-20 show the main results of the calculated changes in surface temperature T_s and atmospheric temperature T_a, depending on the optical thick-

70   2. Climatic Effects of a Nuclear Conflict

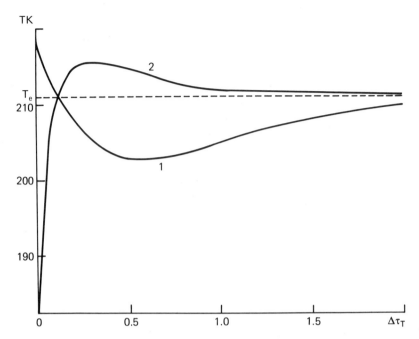

FIGURE 18. The surface atmosphere (2) and temperature (1) of Mars depending on an increase in aerosol optical depth ($\Delta\tau$) in thermal range (which is considered to be four times smaller than in the visible range: $[\tau_v = 4\tau_T]$).

ness τ_T (in the thermal range of the spectrum). Figure 18 presents changes in T_s and T_a for Mars, where the dust absorbs the solar radiation about four times more intensely than the thermal one (Zurek 1982). In the thin Martian atmosphere of carbon dioxide, whose columnar mass is on the average 60 times smaller than that of the Earth's atmosphere (Moroz 1978), the greenhouse effect adds to the value of T_s about 7 K as compared to $T_e = 211$ K [for Mars A = 0.24 and the distance between the Sun and Mars is on the average 1.52 times greater than that from the Sun to the Earth (Pollack 1979)]. With $\tau_T \approx 1.2$, $T_a = T_s = T_e$, the surface temperature drops much below that of the atmosphere and the so-called anti-greenhouse effect takes place, when $T_s < T_e < T_a$ (Ginzburg 1973). This has been confirmed by the results of sounding the Martian atmosphere from the Soviet and American automatic space stations, showing during dust storms the presence of both inversion and isothermal temperature profiles in the planet's atmosphere (Moroz 1978).

Figure 19 presents the results obtained by Golitsyn and Ginzburg (1983, 1985) of changes in T_a and T_s with the injection of dust and smoke into the atmosphere, depending on the optical depth of the radiation attenuation τ_T in the range. For dust, it is assumed that $4\tau_T = \tau_v$ and for smoke $10\tau_T = \tau_v$. Curves 1-3 refer to the dust cloud: curve 1 describes the land surface temperature T_s,

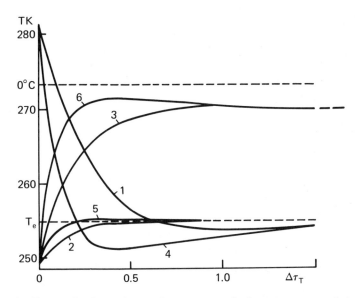

FIGURE 19. Changes in the surface and mean atmospheric temperatures of the Earth depending on an increase in the aerosol optical depth ($\Delta\tau_T$) in the thermal range. 1 = Surface temperature; 2 = atmospheric temperature above land; 3 = atmospheric temperature above the ocean with $\Delta\tau_T/\Delta\tau_v = \frac{1}{4}$ (dust); 4–6 = Earth's surface temperature (4) and atmospheric temperature above land (5) and above the ocean (6) for smoke aerosol with $\Delta\tau_T/\Delta\tau_v = 1/10$ and albedo $A = 0.3$.

curve 2, the atmospheric temperature above land, and curve 3, the atmospheric temperature above the ocean, the temperature changes of which being neglected here.[3] Curves 4–6 describe changes in the Earth's surface temperature (4), the atmospheric temperature above land (5), and above the ocean (6) with the invariable albedo of the system $A = 0.3$ for the smoke-borne aerosols. It can be seen in Fig. 19 that aerosols of dust and smoke origin change the temperature in different ways. In the case of the dust aerosol (curves 1–3), whose absorptivity is lower, the atmospheric temperature changes much more slowly with increasing infrared opacity than in the case of the smoke aerosol that is strongly absorbent. Figure 19 shows that the difference in atmospheric temperature over the ocean and over land may reach 15°C, which can be very important for the dynamics of the atmosphere. This occurs because the ocean under the aerosol cloud becomes cold very slowly due to its large thermal inertia, and its thermal emission serves as an additional source for heating the smoke-laden atmosphere.

Figure 20 presents the curves for the temperature of the land surface (1) and of the atmosphere above land (2) and above the ocean (3) for the smoke

---

[3]The calculations carried out by Pollack et al. (1983) have shown that in 6 months, the ocean surface temperature decreases by about 3° to 5°C.

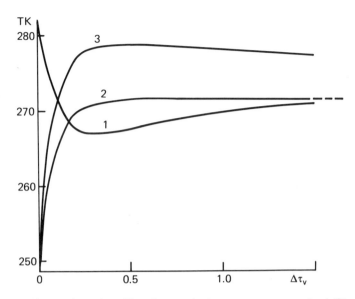

FIGURE 20. Changes in surface *(1)* and atmospheric temperatures over land *(2)* and over the ocean *(3)* with albedo $A(\tau_v) \to 0.1$.

aerosol when the albedo of system A increases with an increase in the optical depth as follows:

$$A(\tau_v) = 0.1 + 0.2 \exp(-1.66\ \tau_v).$$

The formula approximates the albedo course with changing optical depth. In a given case with $A = 0.1$, the limiting temperature T_e is equal to 271.5 K. As can be seen in Figures 18–20, the greatest temperature changes take place with the visual optical depth of an order of 1, since in this case the solar radiation flux that reaches the surface changes most dramatically.

Considerable changes in the thermal regime of the atmosphere and the land surface should alter both the dynamic regime of the atmosphere and the entire hydrological cycle appreciably. Let us consider first a simpler case referring to Mars, the planet that has no oceans, and where dynamic changes have actually been observed during dust storms (Ryan and Henry 1979). The heating of the atmosphere and the cooling of the underlying surface decrease the vertical temperature gradient and even change its sign: instead of falling with height, the temperature starts to rise, which greatly increases its static stability. At the same time there is a weakening of the so-called baroclinic instability that leads to the formation of large cyclonic vortices (Holton 1972). Observations of the Martian surface from the Viking automatic landers have shown that in the Northern Hemisphere, when the atmosphere is clear from dust, there are regular cyclonic occurrences. With the appearance of large dust clouds, cyclonic activity does not

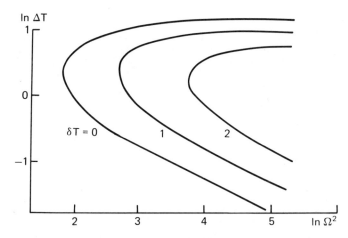

FIGURE 21. Zonal flow stability depending on horizontal temperature difference (ΔT) and angular rotation rate (Ω) with different values of vertical temperature difference (δT). To the left of the curves, the flow is stable; the eddies are formed inside.

occur at all and only tidal fluctuations in atmospheric pressure are observed (Ryan and Henry 1979; Sharman and Ryan 1980).

In regard to the theory of general atmospheric circulation, the Earth and Mars are similar to each other (Golitsyn 1973a). Therefore, the same effects can also be expected to occur in the Earth's atmosphere, only they might be more complex because of the oceanic influence on the atmospheric processes. A simplified theory of the cyclogenesis has been developed and checked in laboratory experiments by Bubnov and Golitsyn (1985). This theory is a generalization of Lorenz's highly truncated model (1962), taking account of the effects of vertical stability on general circulation. Lorenz developed his model to describe the flows in a rotating annular vessel with a cold internal and a warm external wall, which was used earlier by Hide (1958) for modeling atmospheric general circulation (see also Dolzhansky and Golitsyn 1977; Gledzer et al. 1981). The cold internal part simulates the polar regions, and the warm external one the tropical regions. The flows between the walls reproduce well the principle features of the atmospheric general circulation in middle latitudes: the zonal flow and the vortices appearing on it are cyclones and anticyclones.

The nature of the flow in the vessel depends on the temperature difference between the walls, and the rotation rate. There are two basic regimes of the flow which are separated by a curve in the stability diagram of an anvil form (Fig. 21). There is a certain rotation rate Ω_{min} before it is reached, the trajectory of a fluid particle, whatever temperature difference between the walls, looks like a regular spiral, beginning at the warm wall and ending at the cold one: in this way heat exchange between the walls takes place. If the vessel were not rotating, the particle would have moved in the meridional plane, but the Coriolis force drives the particle out of this plane, creating an azimuth (zonal) component in the flow.

74   2. Climatic Effects of a Nuclear Conflict

FIGURE 22. An example of the flow with three cyclonic and three anticyclonic eddies in a circular-shaped vessel rotating clockwise; the inner side of the vessel is cold and the external side is warm. The currents simulate the mid-latitude circulation. With application of a sufficiently strong vertical temperature gradient, the eddies disappear, i.e., the flow becomes stable. (The picture is courtesy of B.M. Bubnov.)

This is the so-called Hadley regime. With an increase in Ω, i.e., when the Coriolis force increases, with $\Omega < \Omega_{min}$ the spiral becomes more and more twisted until with $\Omega = \Omega_{min}$ instability of the flow occurs, the so-called baroclinic instability, when the trajectory of the particle starts to touch the wall two or more times, thus maintaining the heat exchange between the walls. The vortices of cyclonic sign (i.e., twisted in the direction of the vessel's main external rotation) are formed in the liquid. The regime where the vortices are formed is called in meteorology the Rossby regime (Lorenz 1967). Figure 22 represents a photograph of the vessel with three cyclonic and three anticyclonic vortices that can be seen after the injection of aluminum powder into the vessel. With very great temperature differences, the motion again becomes stable, since convection overcomes instability.

If this flow is then superimposed by temperature difference in the vertical direction, so that the higher temperature is above, the structure of the flow changes. According to the calculations carried out by Bubnov and Golitsyn (1985), the stability curve, i.e., the entire "anvil", shifts to the right and down (see Fig. 21). That is indicative of a general increase in flow stability. Laboratory experiments described by Bubnov and Golitsyn have fully confirmed these conclusions.

With regard to the Martian atmosphere during dust storms, this study ascribes the disappearance of the cyclones to an increased atmospheric stability due to a changed sign of the vertical temperature gradient in the dust-laden Martian atmosphere, when the solar radiation is mainly absorbed by dust and the planet's surface becomes cold (see Fig. 18).

This effect of the suppression of the cyclogenesis can also be expected to occur in the smoke-laden Earth's atmosphere, at least where its static stability increases. It is well known that cyclones have neither ever penetrated the stratosphere, nor have they been formed there, because the Earth's atmosphere above the tropopause has a very stable stratification. Therefore, in smoke-laden atmosphere cyclones can be either completely suppressed, or very weak cyclones can form, occupying the layer where the vertical temperature gradient has not yet increased because of the absorption of solar radiation by the smoke.

The general circulation models described by Pittock et al. (1986) show rapid changes in the structure of the general circulation after the introduction of absorbent smoke into the atmosphere. In the normal atmosphere (Lorenz 1967), if the wind is averaged by the longitude, in the meridional plane two or three cells are found in each hemisphere, in the tropical one the so-called Hadley cell, with air ascending close to the equator and descending in subtropical latitudes (where the Coriolis force leads to the formation of trade winds), and in the mid-latitudes the so-called Ferrel cell, when the air rises in high latitudes and descends in the latitudinal belt of 30° to 40°. Here the westerlies prevail. In winter, a cell also appears in polar latitudes. In smoke-filled atmosphere heated by solar radiation, one large Hadley cell is observed with a descending branch in the Southern Hemisphere, which entrains the smoke at the upper levels (Stenchikov 1985a, b).

The models have shown that in smoke-laden atmosphere cloudiness and precipitation are considerably decreased and the hydrological cycle is in general less intense. The idea of the attenuation of the hydrological cycle in smoke-filled atmosphere was first advanced in the USSR at a time when work concerned with the atmospheric consequences of a nuclear war was just commencing (Golitsyn 1983; Golitsyn and Ginzburg 1983; Obukhov and Golitsyn 1983). This attenuation can be explained by the following reasons:

1. Other things being equal, the relative humidity decreases more in the heated atmosphere than in the normal one, which lowers the probability of water vapor condensation.
2. The higher stability of the atmosphere due to increasing vertical temperature gradient decreases the intensity of water exchange between the atmosphere and the Earth's surface where the water sources are found.
3. An increase in the number of particles that can form condensation centers leads to the formation of a large number of fine droplets, which lowers the so-called precipitation potential. It is known that in relatively clear air over

the ocean, the concentration of droplets in clouds is an order of magnitude lower and the droplets themselves are larger than those in clouds over land (Rodgers 1978).
4. The cold continents and warm oceans induce monsoon circulations, but this occurs as a winter monsoon that does not carry moisture to the continents.
5. A decrease in solar energy income to the Earth's surface provokes a fast change in the entire energy balance, which results in an appreciable decrease in the moisture flux to the atmosphere (see above).

Of course, the intrusion of cold air from the continents into the oceanic regions while the ocean is still warm will add greatly to the atmospheric instability, as is observed over the Gulf Stream and Kuroshio. However, since the air temperature over the ocean will be higher on the average than over the continents (see Figs. 19–20), the cold air will be found only in the lower atmospheric layers above the ocean, inducing the formation of not very high mesoscale weather systems, and the precipitation will be mainly formed above the ocean itself (Molenkamp 1985). The coastal regions might experience rapid weather changes during which the warm air from the ocean will raise the temperature near the Earth's surface and the cold continental masses will cause it to decrease significantly. On the whole, of course, the ocean will moderate the cooling effect. The circulation models incorporating ocean-atmosphere interactions show on the average for the hemisphere a temperature decrease that is approximately half that in the interior of the continents.

A decrease in atmospheric precipitation leads to longer residence of aerosol particles in the atmosphere. We have already given the example of the Arctic haze, when light precipitation in winter and spring allows the particles to stay in the atmosphere for several months. All these effects, therefore, can prolong the climatic consequences of smoke-laden atmosphere (US National Research Council 1985; Pittock et al. 1986).

2.2 Other Atmospheric Effects of a Nuclear War

Investigations concerned with estimating the consequences of a nuclear war were initiated by studying the possible destruction of the ozone layer. The American scientists Foley and Ruderman (1973) were the first to note that high-yield nuclear explosions generate fireballs, in which a great number of nitrogen oxides (NO) are formed, namely about 10^{32} NO molecules per 1 Mt of explosive yield. While the mushroom-shaped cloud rises into the stratosphere, the NO molecules interact with the ozone and destroy it; however, the NO molecules act as a catalyst, their number remaining invariable. Different calculations (US National Academy of Sciences 1975, Crutzen and Birks 1982, Izrael et al. 1983) have shown that in a nuclear conflict involving a great number of 1-megaton or larger explosions, about half (to 70%) the stratospheric ozone of the Northern Hemisphere might be destroyed. This would lead to a considerable ozone redistribution by height, because the destruction would mainly occur at a height to which cloud rises (from 10 to 40 km, depending on the explosion yield). At lower altitudes the ozone would be preserved. Such ozone redistribution causes appreciable changes in temperature distribution in the stratosphere and upper troposphere because of the different absorption of ultraviolet radiation in the troposphere and stratosphere. Maximum ozone depletion would take place within about half a year after the explosions, and the regaining of standard ozone levels would continue for 2 to 3 years. After the dissipation of the smoke clouds, the ultraviolet radiation flux will be noticeably enhanced at the Earth's surface, particularly in the spectral band of 0.27–0.3 μm, which could lead to significant biological consequences (a higher mutation level, attenuation of photosynthetic and immune activity, greater incidence of skin cancers and blindness, etc.).

However, comparatively low-yield explosions (tactic weapons) generate fireballs which persist mainly in the troposphere (Crutzen and Birks 1982; Pittock et al. 1986), where, on the contrary, photochemical reactions with participation of nitrogen oxides lead to certain increases in ozone concentration. At the same time, smoke loading of the atmosphere causing reduced sunlight intensity can greatly retard these reactions (Pittock et al. 1986). Ozone and other gases, such as carbon monoxide and dioxide, methane, and ethane can form smog, so that the dissipation of the smoke can be followed by the formation of smog.

During fires and emergencies in fields of natural gas deposits, the so-called greenhouse gases – carbon dioxide, methane, and tropospheric ozone are formed and released. An additional rise in the concentrations of some of these gases (for instance, of methane) will be significant compared to their present standard atmospheric levels (for example, an increase of 70 ppb against the background content of 1650 ppb, see US National Research Council 1985). In the presence of light and nitrogen oxides acting as a catalyst, tropospheric ozone is rapidly produced from methane and oxygen. Its concentration is expected to increase to a maximum of 160 ppb against the background level of 30 ppb (Izrael 1983a, b).

In the event of destruction of fields of natural gas deposits, considerable amounts of nonburnt methane, ethane, and propane would be released. The ethane concentration would reach 50 to 100 ppb (against the background level of 1 to 2 ppb). These gases could enhance the infrared opacity of the atmosphere and, after dissipation of the smoke, raise the Earth's mean surface temperature by several degrees, due to the greenhouse effect (Izrael 1983a, b). At the same time, the nitrogen oxides in the stratosphere, in particular N_2O, introduced into the stratosphere after the nuclear explosions, could arrest a portion of the solar radiation, thereby heating the stratosphere and cooling the Earth's surface by several degrees (Izrael et al. 1984).

At the same time, we know that because of the large thermal inertia of the ocean, the immediate introduction of greenhouse gases into the atmosphere would have the greatest effect on climate only in a few decades (Bryan et al. 1982; Schlesinger et al. 1985). Since the residence time of all the gases mentioned above, except CO_2, does not exceed a few years (Crutzen and Andreae 1984), the enhanced greenhouse effect might not occur in the post-nuclear exchange atmosphere.

The chemical processes in the atmosphere after a nuclear conflict have not yet been studied sufficiently. For instance, according to the estimates from the report by the US National Research Council (1985), nuclear fires can release about 10^{16} g (10^4 Mt) of CO_2 into the atmosphere, which is approximately half the annual amount of CO_2 released into the atmosphere through the combustion of fossil fuels. Since the atmosphere contains 3×10^{18} t of CO_2, the direct effect of this CO_2 release will be insignificant. However, the atmospheric CO_2 content could change as a result of indirect perturbations of the global biospheric carbon cycle, for example as a result of the dying of plants, which could appreciably increase the CO_2 concentration in the atmosphere. This problem has still been very little treated.

Similar estimates can also be cited for methane. According to the US National Research Council Report (1985), about 5×10^{13} g of methane can be introduced into the atmosphere during nuclear fires, which increases its atmospheric amount by about 2% (3×10^{15} g), so that the direct effect of such an increase will not be strong. In a similar way, the direct introduction of water vapor into the atmosphere during nuclear explosions and fires cannot be of great importance for long-term effects, not only because the maximum increased water vapor content is, according to the NRC report of 1985, 1.4×10^{16} g, which is very low compared with its standard content of the order of 10^{19} g, but also because the lifetime of water vapor in the standard atmosphere is only about 10 days. The local perturbations of the atmosphere as a result of the introduction of a considerable amount of water vapor, atmospheric redistribution due to explosions and rising fluxes of warm humid air could be quite significant, but would not change these general conclusions.

The change in the albedo of the land surface, A_s, after a nuclear conflict is of potential importance. Since it is not likely that the fires would cover an area of more than 2×10^6 km², which is less than 1% of the entire Northern Hemisphere area, direct changes in global radiation balance would not be large. Indeed, with an average $A_s \approx 0.2$, even if A_s approaches zero over a 1% area, the average hemispheric surface albedo would decrease by 0.002, which could lower the temperature by no more than several tenths of a degree. The effect of snow pollution in high latitudes (Warren and Wiscombe 1985) by settling smoke and soot particles could well prove to be more pronounced. The snow albedo could decrease from 0.8–0.7 to 0.5–0.25, which would cause snow and sea ice to melt earlier than usual. This would evoke the feedback between the albedo and temperature of the Earth's surface (Budyko 1968), which in turn would result in a warming. Estimation of such a warming has, however, shown that it will be much weaker in absolute value than the cooling produced by the accumulation of aerosol particles in the atmosphere.

However, it is also possible that long-term feedbacks would start to operate in the climatic system, leading to quite opposite effects. With the help of the energy-balance climate model of Robock (1984), and considering the influence of snow and ice on the upper oceanic layer, the integration was carried out for several years after the initial attenuation of solar radiation at the surface (which would be restored to its normal values within a few months). Because of the nonlinearity and inertia of the feedback between temperature and the Earth's surface albedo, in model calculations the intensity of the annual cycle was decreased, which yielded lower summer temperatures. Therefore, smaller amounts of snow and ice melted, which led to a general temperature decrease that continued for several years, with diminishing amplitude. This effect should be studied on the basis of the more detailed climatic models. Most of these models have until now been run for a term of only about a month. In his model, Stenchikov (1985a) integrated for a year but used fixed mean annual insolation conditions, i.e., neglected the annual variations.

On the whole, the long-term climatic consequences of a nuclear war are not yet understood properly. To clarify this problem, it is necessary to use widely differing models, incorporating also seasonal changes in climatic conditions.

One of the present authors has published work on a phenomenon that has until now been studied insufficiently, the effects of radioactive products from explosions on the electrical properties of the atmosphere. The concentration of radioactive products (up to 10^{12} C in one week after the explosions, assuming that 20% of explosive energy is emitted through fission reactions) that are evenly distributed in the troposphere, comprising possibly about 10% of all radioactive products, will be equal to 5×10^{-7} C m^{-3}, i.e., equivalent to an energy emission of about 5×10^{-7} MeV/(m^3 s). This would lead to an ionization of the air that would by several hundred times exceed the ionization that induces a 10% change in the electric conductivity of the air (Izrael et al. 1982; Izrael 1983b, 1984, 1985). It is clear that this can influence the process of cloud formation.

On the whole, it can be concluded that of all the potential effects of a nuclear exchange on atmospheric processes, the most disastrous for the existence of the biosphere and mankind is aerosol cooling. It should, however, be emphasized that this does not exclude other forms of nuclear impact on natural conditions of our planet, which might be dangerous, but are as yet not investigated.

2.3 The Reliability of the Results

THE RELIABILITY OF THE RESULTS OBTAINED

In a number of studies on the climatic effects of a nuclear conflict published in recent years, two problems have been considered, namely the broadening of information on the release of the optically active particles produced during nuclear explosions and fires, and the application of model calculations to obtain information on the distribution of the expected climatic changes (mostly air temperature changes) in space and time.

Solving these problems is fraught with certain difficulties, as has been discussed in a number of works, including those of the present authors (Izrael 1984;

Budyko 1985; Golitsyn 1985). Great attention has been paid to this issue by ENUWAR (Pittock et al. 1986).

The accurate estimation of aerosol amount produced by a large-scale nuclear exchange is possible only by the extensive use of empirical information, particularly on the production of optically active particles in massive fires, when exceptional, as yet unstudied, atmospheric processes could arise, which would influence the formation and distribution of aerosol particles. It is probable that experiments on small-scale objects, compared to those which might be set on fire during a nuclear conflict, will not give accurate information on this problem.

No less difficult are the problems of applying the available climatic models to calculate the highly transient processes leading to climatic changes, which, if they indeed occurred in the geological past, have never been observed by man. The most advanced climatic models give a satisfactory description of the spatial distribution of some meteorological elements under a stationary climatic system, or with comparatively slow changes in the factors affecting this system. These models have been further developed and verified extensively by observational data from the world network of meteorological stations that has been in operation for more than 100 years.

It is, however, very doubtful if the available models are capable of producing very reliable results to describe in detail the development of atmospheric processes during a nuclear exchange. It is, moreover, not even likely that climatic models which would serve this purpose can be developed in the near future.

This latter remark arises not only from general considerations. There is sufficient empirical information on aerosol "microcatastrophes", which are produced by every explosive volcanic eruption. It has already been mentioned that such eruptions are followed by mean air temperature decrease in the hemisphere where the volcano is situated, which continues for a period from several months to 3 years. However, empirical studies have also revealed a complex space and time pattern of air temperature anomalies and a temperature increase in some areas and certain seasons. In this case there is the possibility that, over some limited areas and for a shorter time, changes in meteorological regime are determined not by regular climatic changes induced by a volcanic eruption but by relatively short-term synoptic processes, which are to a considerable (unknown) extent nonpredictable due to the limited stability of the general circulation of the atmosphere and the ocean.

It is no accident that at present there is no realistic theory describing the details of regional changes in the meteorological regime after any single explosive volcanic eruption. At the same time, the development of such a theory by no means ensures its ability to predict much greater changes in time and space in the meteorological regime after a nuclear war.

In addition to these important limitations of climatic models in estimating regional consequences of a nuclear conflict, other causes also lower the accuracy of model calculations.

The spatial resolution of these models is always limited, i.e., these models cannot be applied to describe the processes of the so-called subgrid scale. Even the most elaborate model of the European Medium-Range Weather Forecast Center that includes 18 vertical levels and a space step of 90 km does not simulate thunderstorm systems and many meso-scale processes. In spite of significant improvements in the quality and terms of weather forecasting, even this model cannot make predictions for a period exceeding 10 days. The microphysics of clouds and the processes of removing aerosol from the atmosphere and its microphysical transformations have still been insufficiently studied. No model is capable of simulating the meso-scale processes which in smoke-laden atmosphere might exert a great influence on smoke distribution. This provoked the appearance in 1984 of several publications criticizing the concept of an aerosol climatic catastrophe (Teller 1984; Singer 1984; Barton and Paltridge 1984).

This criticism consisted for the most part in indicating the uncertainties arising from an insufficient knowledge of some of the processes that would occur in the perturbed atmosphere, and to discussing these processes which could moderate climatic effects of a nuclear exchange, such as coagulation, aerosol scavenging, meso-scale processes, etc. However, such criticism was not objective, as it considered only the processes that could reduce the lifetime of aerosols in the atmosphere, and as a rule, presented no reliable quantitative estimates. At the same time there are also processes that considerably increase the residence time of the aerosols, first among them being the attenuation of the hydrological cycle in the atmosphere. In addition, we recall once more the results recently published by Cess et al. (1985). It has been calculated that in the atmosphere with initial smoke optical thickness of $\tau = 3$ in the second decade (days 11-20) after the introduction of smoke into the atmosphere, the precipitation amount for the entire mid-latitudinal belt will decrease by half as compared with the baseline calculation for the standard atmosphere. The dissipation of the majority of the clouds is also noted in the calculations by Malone et al. (1986).

Almost all scientists concerned with the problem of climatic change after a nuclear war support the conclusion of the likelihood of a great aerosol catastrophe. Among the few objections to this point of view are the ideas of Teller (1984), a specialist in the field of atomic physics.

In his article, Teller briefly dwells on the problem of climatic impact of the dust particles and, without any substantiation, concludes that this impact will be comparable with that of a major volcanic eruption, which, according to him, produces appreciable influence on climate, but does not lead to any catastrophic changes.

As has been mentioned in Chapter 1, the greatest volcanic eruptions in the geological past seem to have had catastrophic consequences covering quite extensive territories. Considerable damage was done to animate nature even by the less intense volcanic eruptions of the past. If what Teller means is the climatic impact of less intense explosive volcanic eruptions of the past century, his assertion as to the coincidence of the mass of volcanic aerosol with that of dust particles after a nuclear conflict is quite arbitrary. This assertion is markedly not in agreement with the fact (mentioned by Teller himself) of the insufficient experimental data used in the calculations on the dust mass released into the atmosphere during a nuclear exchange, which should have made him more cautious in his estimates of the dust amount in a post-war atmosphere.

At the same time, there is a physical mechanism that can considerably increase the aerosol mass which would affect the climate after a nuclear war.

Measurements of the amount of dust particles reaching the upper layers of the atmosphere were made during nuclear tests involving single explosions of atomic bombs that did not noticeably change the atmospheric general circulation over vast territories. An entirely different situation would arise with the almost simultaneous explosion of a great number of nuclear warheads. According to the majority of scientists, in this case a dense dust and smoke veil, almost opaque for short-wave radiation, would form and persist for some time. This veil would be lifted by the upward air fluxes to a considerable height, which would mix the tropospheric and the lower stratospheric layers into one single system.

Under such conditions, a considerable portion of aerosol particles would penetrate the lower layers of the stratosphere, some of the aerosols persisting there even after the smoke veil thinned as a result of the deposition of the largest particles, their coagulation, and the attenuation of the influx of aerosol particles from the lower air layers.

In addition to this way of introduction of aerosol particles into the stratosphere, there could be other physical mechanisms inducing changes in the general circulation of the atmosphere during nuclear explosions, which would lead to an increase in aerosol load in the lower stratosphere. After the rupture of the aerosol veil, the physical properties of the modern stratosphere would be restored and the aerosol particles found there would persist in the higher layers of the atmosphere for a long time due to the weak vertical air motion in these layers. It is clear that the available empirical data are insufficient to estimate these processes of accumulation of aerosol particles in the stratosphere.

In more detail, Teller treats the problem of aerosol particles released into the atmosphere as a result of fires set off by nuclear explosions. He thinks that a considerable fraction of smoke particles would be absorbed by the droplets of liquid water found in the atmosphere. However, he does not deny the possibility that an appreciable mass of these particles could reach the higher layers of the atmosphere, where they could stay for a long time.

Teller also indicates some other possible causes of uncertainties in the calculations carried out by Turco et al. (1983) and Covey et al. (1984). Some of these causes are evidently quite correct (for example, those relating to the hypothetical nature of the scale of a possible nuclear conflict), others are clearly one-sided and lead to a selection of those arguments that represent the calculated results as overestimated.

The strongest objections are thus raised less by Teller's actual discussion of details, but by the conclusion to which he comes.

Teller thinks that the probable decrease in the surface air temperature after a nuclear exchange would be smaller than that given by Turco et al. He assumes that the temperature in the mid-latitudes of the Northern Hemisphere might drop by $5°-6°C$. Such a decrease in temperature would substantially ruin crop yields and cause famine. He calls such a prospect "frightful." To avoid it, according to Teller, the USA should store large quantities of foodstuffs.

At the end of his paper, Teller contends that the conclusions about the probability of more or less considerable climate change after a nuclear war, possibly even resulting in the destruction of the biosphere, damage scientific prestige and should not be quoted for political decisions.

The errors that led Teller to such declarations are elementary. Having made a few comments on the high inaccuracy in the calculations he is discussing, Teller concludes that these calculations overestimated the expected temperature decrease by about ten times. However, he misses the fact that at the modern state of research, it is equally easy to prove not an overestimation but an underestimation. In this connection it is clear that Teller's estimate of the cooling in middle latitudes, which evidently refers to the continents, is quite arbitrary. It is difficult to understand how on the basis of such an estimate Teller could give recommendations about definite political decisions (storage of food).

To solve this issue it must be clearly understood that the methods available for determining the impending climatic change after a nuclear war permit prediction of only a sign and order of magnitude for future cooling.

This means that the prediction of the ecological consequences of a nuclear conflict falls only into the rank of a probability. Considering a series of the values, from greatest to smallest, of the expected temperature decrease, we can list here several possible versions of such a prediction, which will include:

1. the destruction of the biosphere;
2. the extinction of many species of animals, plants, and human beings;
3. the extinction of some of the living organisms and the survival of a part of mankind;
4. the absence of serious ecological consequences over the territory not directly involved in military operations.

The latter version seems to have no supporters at present, even Teller considering that such a possibility is very low.

As far as can be understood, he considers the third version to be the only plausible one, hence his recommendations to take certain precautions, including the storage of food in quantities comparable with the annual production. This last idea of Teller makes a painful impression: it is quite conceivable that the economic potential of the developing countries that comprise the greater part of the world's population does not suffice to make such provisions. Thus the realization of such an idea does not stave off the threat of the annihilation of billions of people, in the case of a nuclear war, who would have taken no part in this war themselves.

However, the most important fact lies elsewhere.

As mentioned above, the probable error in the calculations of expected temperature decrease after a nuclear exchange could be no less than of one order of magnitude. The most probable estimates contend that the temperature decrease for the continents would be from several to several tens of degrees, depending on the scale of the conflict. Together with other specialists in the field of atmospheric sciences, who are concerned with this problem, we think that the most probable estimates are those approaching the upper limit. At the same time, there is some probability (perhaps comparatively low) that much lower estimates could also be correct. In addition, there is also a not very high probability that even the highest estimates of the predicted cooling are considerably underestimated (if the albedo were to increase for any reason).

Thus, at present it is necessary to take account of the first three ecological predictions of consequences of a nuclear war listed above. Although Teller seems to consider it possible to wage a nuclear war, basing on the third version of the prediction, this issue loses its significance if the second or even the first versions are not excluded. With such a probability, the idea of a nuclear war is totally lacking in common sense.

Not to follow the example of Teller, who criticizes the studies he discusses with considerable sharpness, we restrict ourselves to saying that Teller's ideas are inappropriate for consideration when taking political decisions.

Now we turn to the aspects of the problem that require further advances in research. It is evident that in modeling climatic consequences, some effects have not been taken into account which can either attenuate or intensify these consequences. What can be considered to be established fact, and what still needs to be investigated?

Let us present here the conclusions of the joint report submitted by Golitsyn and Phillips at the 6th Session of the Joint Scientific Committee concerned with World Climate Research Program, which was also presented at the 37th Session of the WMO Executive Council (Geneva, June, 1985).

Golitsyn and Phillips (1986) noted in their report that scientific problems of a "nuclear winter," leaving aside its biological and chemical consequences, can be divided into two groups:

1. The quantitative refinement of estimates of the smoke mass which would be produced during a concrete nuclear exchange; the determination of the initial properties of the smoke and its distribution at the end of this exchange.
2. The calculation of the resultant aerosol effect on the atmosphere, land and water surfaces, including the atmospheric effect on the smoke.

It has been stated in the joint report that both these groups present great uncertainties. As an example, we recall the wide range of estimated smoke amount $(0.2 \div 6.5) \times 10^{14}$ g, which it is suggested must be taken into account together with the baseline estimate of 1.8×10^{14} g, in the NRC report of 1985. It is important to emphasize that these uncertainties are not one-sided, but either increase or decrease climatic effects.

Earlier studies assumed smoke volume to be $(1 \div 2) \times 10^{14}$ g. This amount is enough to exclude practically any sunlight during the first weeks following a nuclear exchange. These conditions are so extreme that at present the following conclusion seems to be justified.

Prediction of drastic temperature changes within the weeks following the formation of $(1 \div 2) \times 10^{14}$ g of smoke from nuclear fires will not change, except for some details, independently of the success in resolving many uncertainties in the calculations of atmospheric effects included in problems of the second group.

Therefore, reducing uncertainties in studying the problems of the second group is important first of all in order to estimate the effects of a nuclear exchange, which would produce a much smaller amount of smoke, the effects in the southern latitudes, and particularly long-term effects (of the order of 1 year for instance).

Among the uncertainties specified in the joint report, some have already been discussed in this book. They include insufficient knowledge about (a) the dispersion of the smoke cloud (i.e., data on the height that could be reached by the smoke and its extension by atmospheric motions of different scale), (b) the optical aging of the aerosol (i.e., coagulation and chemical evolution leading to modifications of optical properties), and (c) the scavenging of the aerosol by precipitation.

To overcome these uncertainties, such highly elaborated models might be needed as cannot be developed for many years to come. However, since a limited nuclear exchange seems to be an impossibility and a large-scale nuclear war would produce smoke to the order of 10^{14} g (100 Mt) or even more according to the most cautious estimates such as those in the NRC report of 1985, a rapid attenuation of the sunlight flux reaching the Earth's surface would be expected, and as a result of this, the continental temperature might drop by several tens of degrees. These conclusions are the most grave argument for the inadmissibility of a nuclear war.

CONCLUSIONS

Although great reserves of nuclear arms have already existed for several decades, it has only recently been revealed that the unrestricted use of nuclear weapons could lead to a global ecological catastrophe. This shows that the environmental sciences are lagging far behind the development of weapons technology, which in this very case could lead to disastrous consequences.

It can be concluded from this example that the use of nuclear arms can have other extremely dangerous consequences, but the same can be applied to other modern means of mass extermination, that potentially menace the annihilation of mankind because their consequences are still unknown. There is no doubt that this threat is growing rapidly with further progress in military technique.

Two ways may be suggested to reduce the probability of the unpremeditated self-destruction of mankind. The first means a considerable expansion in the investigation of the ecological consequences of the use of modern weapons. The necessity of promoting such studies is beyond all doubt.

At the same time, much more reliable results might be obtained by a second way, leading to a gradual reduction and final complete elimination of all types of weapons that threaten the fate of mankind.

References

Alexandrov VV, Stenchikov GL (1983) On the modelling of climatic consequences of the nuclear war. Proc. Appl. Math. Computing Center of AS USSR, 21 p
Alexandrov VV, Stenchikov GL (1984) On one of the computing experiments modeling the consequences of the nuclear war. J Comp Math Math Phys 24 (No. 1):140-144 (R)[4]
Alexandrov EL et al. (1982) Atmospheric ozone and global climatic changes. Gidrometeoizdat, Leningrad (R)
Alvarez LW et al. (1980) Extraterrestrial cause for the Cretaceous-Tertiary extinctions: Experiment and theory. Science 208:1095-1108
Alvarez W et al. (1982) Current status of the impact theory for the terminal Cretaceous extinction. In: Geological implications of impacts of large asteroids and comets on the Earth. Geol Soc Am Spec Pap (Reg Stud) N 190 pp 305-315
Ambio (1982) vol. 11, no. 2/3
Asaro F et al. (1982) Geochemical anomalies near the Eocene-Oligocene and Permian-Triassic boundaries. In: Geological implications of impacts of large asteroids and comets on the Earth. Geol Soc Am Spec Pap (Reg Stud) No. 190, pp 517-528
Asaturov ML (1977) The formation of stratospheric sulphate aerosol layer. Proc State Hydrol Inst 247:45-54 (R)
Asaturov ML (1979) Modeling of stratospheric aerosol layer. Proc State Hydrol Inst 257:81-88 (R)
Asaturov ML (1981) On the question of the formation of stratospheric aerosol. Proc State Hydrol Inst 271:113-122 (R)
Asaturov ML (1984a) The model of aerosol layer formation in the stratosphere. Meteorol Hydrol No. 2, pp 31-38.
Asaturov ML (1984b) Aerosol evolution after large ejections into the stratosphere. Meteorol Hydrol No. 11, pp 59-66 (R)
Barenblatt GI, Golitsyn GS (1974) Local theory of nature dust storms. J Atm Sci 33 (No. 10):1917-1933
Barton I, Paltridge GW (1984) Twilight at noon overstated, vol 13 No. 1. Ambio, pp 49-51. Reply by PJ Crutzen. Darkness after a nuclear war. Ibid, pp 52-54
Bolle HJ (1982) Radiation and energy transport in the Earth-atmosphere system. In: Hutzinger O (ed) Handbook of Environmental Chemistry, vol 1, pt.B. Springer, Berlin, Heidelberg, New York, pp 131-303
Brinkman AW, McGregor J (1983) Solar radiation in dense Saharan aerosol in northern Nigeria. Q J R Meteorol Soc 109:831-847

---

[4] R= in Russian.

Brunswig H (1982) Feuersturm über Hamburg. Motorbuch, Stuttgart, 245S

Bryan K, Komro FG, Manabe S, Spelman MJ (1982) Transient climate response to increasing atmospheric carbon dioxide. Science 215:56-58

Bubnov BM, Golitsyn GS (1985) Theoretical and laboratory modelling of statistical stability effects on the structure of general atmospheric circulation. Dokl AS USSR 281 (No. 5):1076-1079 (R)

Budyko MI (1968) On the origin of Ice Ages. Meteorol Hydrol No. 11, pp 3-12 (R)

Budyko MI (1969). Climatic changes. Gidrometeoizdat, Leningrad, 35 p (R)

Budyko MI (1971) Climate and life. L: Gidrometeoizdat. 472 p (R). (English translation: Miller DH (ed) 1974 Academic Press, London, 470 p)

Budyko MI (1972) Man's impact on climate. Gidrometeoizdat, Leningrad, 47 p (R)

Budyko MI (1974) Climatic change. Gidrometeoizdat, Leningrad, 280 p (R) (English translation: American Geophysical Union, Wash, DC 261 p)

Budyko MI (1980) The Earth's climate: past and future. Gidrometeoizdat, Leningrad, 352 p (R). (English translation: Academic Press, London, 1980, 307 p)

Budyko MI (1982) Changes in the environment and successive faunas. Gidrometeoizdat, Leningrad, 77 p (R)

Budyko MI (1984) The evolution of the biosphere. Gidrometeoizdat, Leningrad, 488 p (R). (English translation: Reidel, Dordrecht, 1985)

Budyko MI (1985) Aerosol climatic catastrophes. Priroda (Nature) No. 6, pp 30-38 (R)

Budyko MI, Vinnikov KY (1983) The problem of detecting anthropogenic changes in global climate. Meteorol Gidrol No. 9, pp 5-13 (R)

Budyko MI, Ronov AB, Yanshin AL (1985) History of the Earth's atmosphere. Gidrometeoizdat, Leningrad, 208 p (R). (English translation: Springer, Berlin Heidelberg New York 1987)

Bull GA (1951) Blue sun and moon. Meteorol Mag 80:1-4

Cadle RD (1972) Composition of the stratospheric sulfate layer. EOC 53:812-820

Carlson TN (1979) Atmospheric turbidity in Saharan dust outbreaks as determined by analysis of satellite brightness data. Mon Wea Rev 107:322-335

Cess RD, Potter GL, Chan SJ, Gates WL (1985) The climatic effects of large injections of atmospheric smoke and dust: a study of climate feedback mechanisms with one- and three-dimensional climate model. Paper presented at SCOPE/ENUWAR Symposium at Hiroshima, February, 1985. Also: Preprint Lawrence Livermore Nat'l laboratory USRL-92504, April, 1985

Charlson RG, Ogren JA (1982) The atmospheric cycle of elemental carbon. In: Particulate carbon atmospheric life cycle. Plenum, New York, pp 3-18

CLIMAP Project Members (1976) The surface of ice-age Earth. Science 191:1131-1137

Colbert EN, Cowels RB, Bogert CM (1946) Temperature tolerances in American alligator and their bearing on the habits, evolution and extinction of the dinosaurs. Bull Am Mus Nat Hist 26(No. 7)

Cotton WR (1985) Atmospheric convection and nuclear winter. Amer Sci 73:275-280

Covey C, Schneider SH, Thompson SL (1984) Global atmospheric effects of massive smoke injections from a nuclear war: Results from general circulation model simulations. Nature 308:21-31

Crutzen PJ, Andreae MO (1984) Atmospheric chemistry. In: Malone TF, Roederer JG (eds) Global change. Cambridge Univ Press, pp 75-113

Crutzen PJ, Birks JW (1982) The atmosphere after a nuclear war: Twilight at noon. Ambio 11(No. 2/3):114-125

Crutzen FJ, Galbally IE, Brühl C (1984) Atmospheric effects from post-nuclear fires. Clim Change 6(No. 3):323-364

Davies RW (1959) Large-scale diffusion from an oil fire. In: Frenkiel FN, Sheppard PA (eds) Atmospheric diffusion and air pollution. Adv in Geophys, Vol. 5. Academic Press, London, pp 413-415

Davitashvili LS (1969) Causes of organisms' extinction. Nauka, Moscow, 440 p (R)

Development and change of organic world at the Mesozoic-Cenozoic boundary. Vertebrates /VN Shimansky (Ed.) 1978. Nauka (Science), Moscow, 136 p (R)

Devlishev PP et al. (1979) Studying possible usage of lazer method in sounding fires. In: Burning and forest fires. Krasnoyarsk, pp 158-164 (R)

Dolzhansky FV, Golitsyn GS (1977) Laboratory modeling of global geophysical streams (Review). Izv AS USSR, Phys Atm Ocean 13(No. 8):795-819 (R)

Ebert CHV (1963) The meteorological factor in the Hamburg fire. Weatherwise 16(No. 2):70-75

Fedorov KN (1984) This capricious infant El-Nino. Priroda (Nature) No. 8, pp 65-74 (R)

Feigelson YM (1970) Radiant heat exchange and clouds. Gidrometeoizdat, Leningrad, 230 p (R)

Feigelson YM et al. (1981) Radiation in cloudy atmosphere. Gidrometeoizdat, Leningrad, 280 p (R)

Fifield R (1983) Mere hiccups in the history of time. New Sci 98:704-706

Foley HM, Ruderman MA (1973) Stratospheric NO_x production from past nuclear explosions. J Geophys Res 78:4441-4449

Frakes LA (1979) Climate throughout geologic time. Elsevier Amsterdam, 310 p

Francis P (1983) Giant volcanic calderas. Sci Am 248(No. 6):60-70

Ganapathy R (1980) A major meteorite impact on the Earth 65 million years ago: evidence from Cretaceous-Tertiary boundary clay. Science 209:921-923

Ganapathy R (1982) Evidence for a major meteorite impact on the Earth 34 million years ago: implication on the origin of North American tektites and Eocene extinction. In: Geological implications of impacts of large asteroids and comets on the Earth. Geological Society of America Special Paper No. 190, pp 513-516

Ginzburg AS (1973) On the radiation regime of the Martian surface and dust atmosphere. Dokl AS USSR 208(No. 2):295-298 (R)

Ginzburg AS (1985) "Nuclear winter" is an actual threat to humankind. USA: economics, politics, ideology. Nauka No. 3 Moscow (183), pp 50-59 (R)

Ginzburg AS, Golitsyn GS, Demchenko PF (1985) Development of very turbid convective boundary layer. In: Proc 9th International conference on cloud physics. V. 4. Gidrometeoizdat, Leningrad (R)

Glasstone S, Dolan PJ (Eds) (1977) The effects of nuclear weapons. US Dept of Defense III edn

Gledzer YB, Dolzhansky FV, Obukhov AM (1981) Systems of hydrodynamical type and their usage. Nauka, Moscow, 366 p (R)

Golitsyn GS (1973a) Introduction to the dynamics of planetary atmospheres. Gidrometeoizdat, Leningrad, 104 p (R)

Golitsyn GS (1973b) On the Martian dust storms. Icarus 18(No. 1):113-119

Golitsyn GS (1983) Report at the All-Union Conference of scientists for deliverance of humankind from the nuclear war threat, for disarmament and peace. Vestn AS USSR, No. 9, pp 57-60 (R)

Golitsyn GS (1985) Aftereffects of the nuclear war for the atmosphere. Priroda (Nature), No. 6, pp 22-29 (R)

Golitsyn GS, Ginzburg AS (1983) Climatic aftereffects of the nuclear conflict and some natural analogues. Scientific investigation of the Committee of Soviet scientists for peace against the nuclear threat, Moscow, 21 p (R)

Golitsyn GS, Ginzburg AS (1985) Comparative estimates of climatic consequences of Martian dust storms and of possible nuclear war. Tellus 37B:177-183

Golitsyn GS, Phillips NA (1985) Possible climatic consequences of a major nuclear war. Report to XXXIX Executive Council of the World Meteorological Organization, Geneva, June 1985. WCP-113

Goody PM (1964) Atmospheric radiation: Theoretical basis. Oxford Univ Press, London

Gostintsev YA et al. (1985) The turbulence thermals in the stratified atmosphere. Preprint Inst Chemical Physics, AS USSR, 46 p (R)

Grigoriev AA, Lipatov VB (1978) Smoke pollution of the atmosphere according to observations from space. Gidrometeoizdat, Leningrad, 48 p (R)

Hammer EU, Clausen HB, Dansgaard W (1980) Greenland ice sheet evidence of postglacial volcanism and its climatic impact. Nature 288:230-235

Hansen JE, et al. (1981) Climatic impact of increasing atmospheric carbon dioxide. Science 213:957-966

Harwell MA, Hutchinson TC (1986) Environmental consequences of nuclear war, Vol 2. Biological and Agricultural Consequences SCOPE 28. Wiley, Chichester, UK

Herman G (1981) Causes of massive biotic extinctions and explosive evolutionary diversification throughout Phanerozoic time. Geology (Boulder) 9:104-108

Hide R (1958) Some experiments on thermal convection in a rotating liquid. Q J R Meteorol Soc 79(No. 339):161-180

Hobbs PV et al. (1982) Particles and gases in the emissions from the 1980-1981 volcanic eruptions of Mt. St. Helens. J Geophys Res 87:11062-11086

Holton JR (1972) An introduction into dynamic meteorology. Academic Press, New York, London, 319 p

Hsü KJ (1980) Terrestrial catastrophe caused by cometary impact at the end of the Cretaceous. Nature 285:201-203

Hsü KJ et al. (1982) Mass mortality and its environmental and evolutionary consequences. Science 216:249-256

Humphreys WJ (1940) Physics of the air. New York: McGraw Hill

International Assessment of the Impact of an Increased Atmospheric Concentration of Carbon Dioxide on the Environment. WMO/ICSU/UNEP Conference, Villach, Austria, October 1985

Izrael YA (1973) Isotopic content of radioactive fallout. Gidrometeoizdat, Leningrad, 160 p (R)

Izrael YA (1983a) Speech at the 9th Congress of WMO (Geneva, May 1983) on the question Meteorology and Society. WMO Issue, 1984 (R)

Izrael YA (1983b) Ecological consequences of possible nuclear war. Meteorol Hydrol No. 10, pp 5-10 (R)

Izrael YA (1984) Ecology and control of the state of natural environment. Gidrometeoizdat, Leningrad, 560 p (R)

Izrael YA (1985) About choosing the major factors in calculating geophysical and ecological aftereffects of possible nuclear war. Dokl AS USSR 281(No. 4):821-825 (R)

Izrael YA et al. (1982) The problem of anthropogenic ejections of krypton-85 into the atmosphere. Meteorol Hydrol No. 6, pp 5-15 (R)

Izrael YA, Petrov VN, Severov DA (1983) On the influence of atmospheric nuclear bursts on ozone content of the atmosphere. Meteorol Hydrol No. 6, pp 5-15 (R)

Izrael YA, Karol IL, Kiselev AA, Rozanov YV (1984) Modeling of changes in the composition and thermal regime of the atmosphere after possible nuclear war. Report at Soviet-American Meeting on small contaminants in the atmosphere. Vilniyus (R)

Jaenicke R (1981) Atmospheric aerosol and global climate. In: Climatic variations and variability: facts and theories. Reidel, Dordrecht, pp 577-597

Junge C (1963) Atmospheric chemistry and radioactivity. Academic Press, London

Kalitin NN (1920) On the question about the time of onset of optical anomaly in 1912. Izv GFO, No. 1, pp 11-17 (R)

Karol IL (1977) Changes in global content of stratospheric aerosols and their connection with fluctuations in mean direct solar radiation and temperature at the Earth's surface. Meteorol Gidrol, No. 3, pp 32-40 (R)

Karol IL (1983) Gas admixtures in the atmosphere and global climatic changes. Meteorol Hydrol, No. 8, pp 108-116 (R)

Karol IL, Pivovarova ZI (1978) The relationship between stratospheric aerosol concentration and solar radiation fluctuations. Meteorol Hydrol, No. 9, pp 35-42

Kastner M, Azaro F, Michel HV, Alvarez W, Alvarez LW (1984) The precursor of the Cretaceous-Tertiary boundary clays at Stevns Klint, Denmark and DSDP Hole 465 A. Science 226 (No. 4671):137-143

Kelly PM, Sear CB (1984) Climatic impact of explosive volcanic eruptions. Nature 311:740-743

Kelly PM et al. (1985) The extended Northern Hemisphere surface air temperature record: 1851-1984. In: Third Conference on Climate Variations and Symposium on Contemporary Climate: 1850-2100. January 8-11, 1985. Los Angeles, California American Meteorological Society, pp 35-36

Kerr JW (1971) Historic fire disasters. Fire Res Abstr Rev 13:1-16

Kerr JW et al. (1971) Nuclear weapons effects in a forest environment. Thermal and Fire. Report No. 2; TP 2-70. Washington, DC: Defense Nuclear Agency

Kimball HH (1918) Volcanic eruptions and solar radiation intensivities. Mon Wea Rev 46(No. 8):355-356

Knox J (1985) Microphysical mesoscale aspects of nuclear winter and new directions in assessments. Preprint UCRL-91359, Lawrence Livermore Nat'l Laboratory

Kondratjev KY (1985) Volcanos and climate. Reviews of science and technology. Ser Meteorol Klimatol VINITI, Moscow 14 (R)

Kondratjev KY, Ivanov VA, Pozdnyakov DV (1984) Natural and anthropogenic aerosols: comparative analysis. Report at a seminar SCOPE/ENUWAR Leningrad, (R)

Kondratjev KY, Baibakov SN, Nickolsky GA (1985) Nuclear war, atmosphere and climate. Nauka v SSSR (Science in the USSR), No. 2, 3 (R)

Korovchenko AS (1958) Meteorological conditions of flights over mountainous and forest areas. Civil aviation No. 10, pp 32-33 (R)

Lack D (1954) The natural regulation of animal numbers. Clarendon, Oxford, 344 p

Lamb HH (1969) Activité volcanique et climat. Revue de géographie physique et de géologie dynamique, Vol XI, No. 3

Larson DA, Small RD (1982) Analysis of the large urban fire environment. I. Theory. II. Parametric analysis and model city simulations. PSR Report 1210. Calif Pacific Sierra Res Corp, Santa Monica

Liou K-N (1980) An introduction to atmospheric radiation. Academic Press, London

Lorenz EN (1962) Simplified dynamic equations applied to the rotating basin experiments. J Atmos Sci 19(No. 1):39-51

Lorenz EN (1967) Nature and theory of the general circulation of the atmosphere. WMO, Geneva

MacCracken MC, Walton J (1984) The effects of interactive transport and scavenging of smoke on the calculated temperature change resulting from large amounts of smoke:

Paper presented at the International Seminar on Nuclear War, 4th Session, Erico, Sicily, Aug. 19-24

Malone RC, Aner LH, Galtzmaier GA, Wood MC, Toon OB (1986) Nuclear winter: three-dimensional simulations including interactive transport, scavenging and solar heating of smoke. J Geophys Res, 91:1039-1053

Manins PC (1985) Cloud heights and stratospheric injections resulting from a thermonuclear war. Atmos Environ v. 19, 1245-1255

Marov MY (1981) Planets of the Solar System. Nauka (Science), Moscow, 265 p (R)

Masaitis VL, et al. (1980) Geology of astroblems. Nedra, Leningrad, 231 p

Masaitis VL, Mashchak MS (1982) Impact events at the Cretaceous-Paleogene boundary. Dokl AS USSR v 265 (No. 6):1500-1503 (R)

Mass C, Robock A (1982) The short-term influence of the Mount St. Helens volcanic eruption on surface temperature in the northwest United States. Mon Wea Rev 110:614-622

Mayr E (1963) Animal species and evolution. Belknap Press of Harvard Univ Press Cambridge, Mass 797 p

McGhee GR (1982) The Frasnian-Fammenian extinction event: a preliminary analysis of Appalachian marine ecosystems. In: Geological implications of impacts of large asteroids and comets on the Earth. Geol Soc Am Spec Pap No. 190, pp 491-500

McLaren DJ (1970) Time, life and boundaries. J Paleontol 44:801-815

McLaren DJ (1982) Frasnian-Fammenian extinctions. Geol Soc Am Spec Pap No. 190, pp 477-484

McLaren DJ (1985a) Mass extinction and iridium anomaly in the Upper Devonian of Western Australia: A commentary. Geology (Boulder) 13:170-172

McLaren DJ (1985b) Ammonoids and extinctions. Nature 313:12-13

Molenkamp CR (1979) An introduction to self-induced rainout. LLNL Report UCRL-52669

Molenkamp CR (1985) Mesoscale simulation of coastal flows during nuclear winter. Paper presented at Symposium M-14 Climate effects of nuclear war. IAMAP/IAPSO Joint Assembly, August 5-16, 1985, Honolulu, Hawaii

Moroz VI (1978) Physics of planet Mars. Nauka (Science), Moscow, 454 p (R)

Newell ND (1967) Revolutions in the history of life. In: Uniformity and Simplicity. Geol Soc Am Spec Pap No. 89, pp 63-91

Newell ND (1980) Now-asteroid-caused extinctions. Science News 117(No. 2):22

Obukhov AM, Golitsyn GS (1983) Possible atmospheric aftereffects of nuclear conflict. Earth and the Universe No. 6, pp 5-13 (R)

Obukhov AM, Golitsyn GS (1984) Nuclear war: effects on the atmosphere. In: World and disarmament. Nauka (Science), Moscow, pp 92-102 (R)

Ogren JA (1982) Deposition of particulate elemental carbon from the atmosphere. In: Particulate Carbon Atmospheric Life Cycle. Plenum, New York, pp 379-391

O'Keefe JD, Ahrens TJ (1982) Impact mechanics of Cretaceous-Tertiary extinction bolide. Nature 298:123-127

Oliver RG (1976) On the response of hemispheric mean temperature to stratospheric dust: an empirical approach. J Appl Meteorol 15(No. 9):333-350

Palmer AR (1982) Biomere boundaries: a possible test for extraterrestrial perturbations of the biosphere. In: Geological implications of impacts of large asteroids and comets on the Earth. Geol Soc Am Spec Pap No. 190, pp 469-475

Patterson EM, Marshall BT, Rahn KA (1982) Radiative properties of the Arctic aerosol. Atmos Environ 16:2967-2977

Penner JE, Haselman LC, Jr (1985) Smoke inputs to climate models: optical properties and height distribution for nuclear winter studies. Preprint UCRL-92523

Peterson KR (1970) An empirical model for estimating world-wide deposition from atmospheric nuclear detonations. Health Phys 18:357-378

Pittock AB, Ackerman TP, Crutzen PJ, MacCracken MC, Shapiro CS, Turco RP (1986) Environmental consequences of nuclear war, Vol 1. Physical, SCOPE 28. Wiley, Chichester, UK

Playford PE, McLaren DJ, Orth CJ, Gilmore JS, Goodfellow WD (1984) Iridium anomaly in the Upper Devonian of the Canning Basin, Western Australia. Science 226:437-439

Pollack JB (1979) Climatic change on the terrestrial planets. Icarus 37(No. 3):479-553

Pollack JB, Toon OB, Khare BN (1973) Optical properties of some terrestrial rocks and glasses. Icarus 19:372-389

Pollack JB, Toon OB, Sagan C (1975) The effect of volcanic activity on climate. Proc WMO/IAMAP Symp on Long-Term Climatic Fluctuations, WMO, Geneva, No. 421, pp 279-285

Pollack JB et al. (1976) Volcanic explosions and climatic change. In: Pollack JB, Toon OB, Sagan C et al. A theoretical assessment. J Geophys Res 81(No. 6):1071-1083

Pollack JB et al. (1983) Environmental effects of an impact-generated dust cloud: implications for the Cretaceous-Tertiary extinctions. Science 219:287-289

Price P (1985) Ideas about the Earth as a dynamic body are improved. Priroda (Nature) No. 3, pp 103-104 (R)

Proceedings of the Soviet-American Meeting on the Study of Climatic Effects of Increased Atmospheric Carbon Dioxide, Leningrad, June 15-20, 1981 (1982) Gidrometeoizdat, Leningrad, 56 p (R)

Rampino MR, Self S (1982) Historic eruptions of Tambora (1815), Krakatau (1883) and Agung (1963): their stratospheric aerosols and climatic impact. Q Res 18:127-143

Raup DM (1979) Size of the Permian-Triassic bottleneck and its evolutionary implications. Science 206:217-218

Raup DM, Sepkosky JJ (1982) Mass extinctions in the marine fossil record. Science 215(No. 4539):1501-1503

Raup DM and Sepkosky JJ (1984) Periodicity of extinctions in the geologic past. Proc Nat Acad Sci, v 81, Febr. 1984.

Robock A (1984) Nuclear winter: snow and ice feedbacks prolong effects. Nature 310: 668-670

Rodgers RR (1978) A short course in cloud physics, Chapter 5, Pergamon, New York

Rosen H, Novakov T (1983) Combustion-generated carbon particles in the Arctic atmosphere. Nature 306:768-778

Royal Society of Canada (1985) Nuclear winter and associated effects. Royal Society of Canada, Ottawa

Russell DA (1979) The enigma of the extinction of the dinosaurs. Annu Rev Earth Planet Sci 7:163-182

Russell DA (1982) The mass extinctions of the late Mesozoic. Sci Am 246(No. 1): 48-65

Rust K (1982) Volcanos and volcanism. Mir, Moscow, 344 p (R)

Ryan JA, Henry RM (1979) Mars atmospheric phenomena during major dust storms as measured at the surface. J Geophys Res 84(No. 6):2821-2829

Safronov MA, Vakurov AD (1981) Forest fire. Nauka (Science), Novosibirsk, 240 p (R)

Savinov SI (1913) The greatest values of solar radiation intensity according to observations in Pavlovsk since 1892. Izv AS USSR, ser 6 Vol 7, No. 12 (R)

Schlesinger ME, Gates WC, Han Y-J (1985) The role of the ocean in the CO_2-induced climate change: preliminary results from the OSU coupled atmosphere-ocean general circulation model-Report No. 60, Climatic Research Institute, Oregon State University, Corvallis 40 p

Seiler W, Crutzen PJ (1980) Estimates of gross and net fluxes of carbon between the biosphere and the atmosphere from biomass burning. Clim Change 2:207-247

Sharman RD, Ryan JA (1980) Mars atmosphere pressure periodicities from Viking observations. J Atmos Sci 37(No. 9):1994-2002

Schmalgauzen II (1940) Ways and laws governing the evolutionary process. Publishing House AS USSR, Moscow-Leningrad, 231 p (R)

Shoemaker EM (1983) Asteroid and comet bombardment of the Earth. Ann Rev Earth Planet Sci 11:461-494

Simon C (1981) Clues in the clay. Science News 120:314-315

Simpson GG (1983) Private communication.

Singer F (1984) Is the "nuclear winter" real? Nature 310:625. (Reply by SL Thompson, SH Schneider, C Covey, Ibid, pp 625-626)

Smit J, Hertogen J (1980) An extraterrestrial event at the Cretaceous-Tertiary boundary. Nature 285:198-200

Smith CD Jr (1950) The wide spread smoke layer from the Canadian forest fires during late September 1950. Mon Wea Rev 78:180-184

Stanley SM (1984) Mass extinctions in the ocean. Sci Am 250(No. 6):146-154

Stenchikov GL (1985a) Possible climatic consequences of nuclear war: ejections and distribution of optically active admixtures in the atmosphere. Proc IXth International Conference on Cloud Physics Vol 4. Gidrometeoizdat, Leningrad (R)

Stenchikov GL (1985b) Mathematical modeling of climate. Priroda (Nature) No. 6, pp 39-50 (R)

Stith JL, Radke LF, Hobbs PV (1981) Particle emissions and the production of ozone and nitrogen oxides from the burning of forest slash. Atmos Environ 15:73-82

Stommel H, Stommel E (1979) The year without a summer. Sci Am 240(No. 6)

Stommel H, Stommel E (1983) Volcano weather: the story of a year without summer. Seven Seas, Boston, 177 p

Stothers RB (1984) Mystery cloud of AD 536. Nature 307:344-345

Teller E (1984) Widespread after-effects of nuclear war. Nature 310:621-624

The night after: climatic and biological consequences of a nuclear war (1985) Mir, Moscow, 166 p Velikhov ER (ed.)

The quest for a catastrophe (1980) (Editorial) Science News 118(No. 9):134

Thompson SL (1985) Global interactive simulations of nuclear war smoke. Nature, 317:35-39

Thomsen DE (1984) Nemesis: searching for the sun's deadly companion star. Science News 126:134

Toon OB et al. (1982) Evolution of an impact-generated dust cloud and its effects on the atmosphere. In: Geological implications of impacts of large asteroids and comets on the Earth. Geol Soc Am Spec Pap No. 190, pp 187-200

Turco RP et al. (1979) A one-dimensional model describing aerosol formation and evolution n the stratosphere. I. Physical processes and mathematical analogs. J Atmos Sci 36(No. 4):699-717

Turco RP, Toon OB, Ackerman T, Pollack JB, Sagan C (1983) Nuclear winter: Global consequences of multiple nuclear explosions. Science 222:1283-1293

Twomey S (1977) Atmospheric Aerosols. Elsevier, New York

Understanding climatic change: a program for action (1975) The report of US Committee of GARP. Natl Acad Sci USA, Washington

Urey HC (1973) Cometary collisions and geological periods. Nature 242:32-33

Valentine JW (1968) Climatic regulation of species diversification and extinction. Geol Soc Am Bull 79:273-276

Warren SG, Wiscombe WJ (1985) Dirty snow after nuclear war. Nature 313:467-470

Watson HH (1952) Alberta forest fire smoke. Weather 7:128-130

US National Academy of Sciences (1975) Long-term world effects of multiple weapons detonations. US National Academy of Science, Washington DC

US National Academy of Sciences (1982) Carbon dioxide and climate: a second assignment. US National Academy of Sciences, Washington DC

US National Research Council (1985) The effects on the atmosphere of a major nuclear exchange. National Academy Press, Washington DC, 193 p

Weisburd S (1984) Sister star scenario: sound or shot? Science News 126:279

Wexler H (1950) The great smoke pall, September 24-30, 1950. Weatherwise, December 1950, pp 129-142

Wolfson NI, Levin LM (1981) Investigation of meteotron jet distribution in clouds conformably to active influence. Proc Inst Appl Geophys 46:50-68 (R)

Yanshin AL (1961) On the depth of salt-producing basins and some questions of the formation of thick salt layers. Geol Geophys No. 1, pp 3-15 (R)

Zeibold Y (1985) It is necessary to establish how quickly geological processes go. Priroda (Nature) No. 3, pp 101-102 (R)

Zurek RW (1982) Martian great dust storms: an update. Icarus 50(No. 2/3):288-310

Index

A
Aerosol, 8, 13, 16–20, 37, 39, 41–47, 51, 58, 62, 63, 69, 71, 76, 78, 80, 81, 84
Albedo, 68, 71, 78
Ammonoidea, 31
Animate nature, 26, 33, 36, 81
Anthropogenic climate catastrophe, 39, 40
Arctic haze, 24, 76
Asteroid, 19, 20, 31, 33, 35, 37, 38, 42, 68, 69
Astroblems, 33
Atmosphere, 3–8, 16, 21, 35, 39, 50, 51, 59, 65–70, 72, 75–81, 84

B
Baroclinic instability, 72, 74
Belemnoidea, 31
Biosphere, 2, 4, 5, 6, 15, 19, 32, 38, 46, 79
Blue sun and moon, 25
Bony fishes, 31
Brachiopodes, 31
Brown diffusion, 61

C
Carbon dioxide, 3, 4, 8, 39, 40, 41, 77, 78
Carboniferous, 1, 6
Celestial bodies, 19, 33
Climatic system, 13, 17, 34, 41, 42, 80
Climatic zone of life, 35

Coagulation, 61, 81, 82, 85
Comets, 33, 37
Concept of catastrophism, 27
Coriolis force, 73, 75
Cretaceous, 2, 20, 29, 30, 31, 34, 35
Critical epochs, 11, 26, 27, 30
Crocodiles, 2
Crossopterygians, 28

D
Delta-Eddington approximation, 62
Devonian, 36
Dinosaurs, 28, 31
Drought, 9

E
Earth-atmosphere system, 3, 69
Ecological catastrophes, 9, 45, 83, 84
El Niño effect, 10
Energy, 20, 33, 34, 65, 68
Extinction coefficient, 63

F
Ferrel cell, 75
Fires, 24, 41, 52, 53, 57, 61, 78–80, 82, 84
Flora and fauna, 1, 11, 31, 35

G
Glacier advance, 32
Greenfield gas, 61

Greenhouse effect, 3, 4, 66–70, 77, 78
Greenhouse gases, 77

H
Hadley cell, 75
Hamburg, 57, 61
Hiroshima, 52

I
Impact craters, 20
Invertebrates, 28, 29, 31

K
K/Ar dating method, 29

L
Lithogenesis, 1
Lithosphere, 37

M
Mars, 22, 23, 64, 69, 70
Mesoscale weather systems, 76
Mesozoic, 2, 3, 35
Meteorites, 19, 33, 53
Microcatastrophes, 80
Mie parameter, 63
Models, 68, 69, 73, 76, 78, 85

N
Nagasaki, 52, 61
Nemesis, 37
Nuclear bursts, 47
Nuclear conflict, 39, 42, 44, 46, 75, 78, 80–82
Nutrition sources, 11, 34

O
Ocean, 16, 20, 34, 67, 68, 69, 71, 72, 76, 77, 80
Oort cloud, 37
Ordovician, 29

Organic evolution, 38
Organisms, 1–5, 9, 10, 11, 17, 19, 26, 28, 29, 32, 35, 36, 83
Ostracodes, 31
Oxygen, 6, 32
Ozone, 76, 77

P
Paleoanthropology, 29
Permian, 1, 6, 29
Phanerozoic, 1, 3, 30, 36, 37
Phase function, 63
Photosynthesis, 2, 20, 21, 34, 35, 44
Planktonic foraminifera, 31
Pleistocene glaciation, 10
Precambrian, 4

Q
Quaternary, 1

R
Radiation balance, 78
Refraction index, 51, 62, 63
Reptiles, 3
Rhynchocephalians, 28
Rossby regime, 74

S
"Scientific revolution," 36
Sea of Darkness, 23
Smoke, 25, 45, 51, 54–60, 62, 65, 69, 75, 81, 84
Solar constant, 3
Solar radiation, 3, 12, 14, 21, 34, 41, 44, 51, 59, 65, 66, 67, 69, 75, 77, 79
Solar system, 4, 35, 37
Stratosphere, 19, 25, 34, 48, 49, 57–59, 75, 77, 82

T
Tactic weapons, 77
Taxonomic groups, 30

Tertiary, 20, 31
Thermal, inertia, 16
Thermals, 56, 57, 61
Trade winds, 23, 75
Triassic, 6, 29
Tropopause, 57, 75
Troposphere, 42, 51, 53, 59, 69, 77

U
Unifermitarianism, 36

V
Vertebrates, 29, 35
Volcanic eruptions, 8, 9, 12, 13, 21, 33, 34, 37, 80, 81

W
White Earth, 4
Würm glaciation, 2, 3

Y
Year without summer, 15